Homework Book

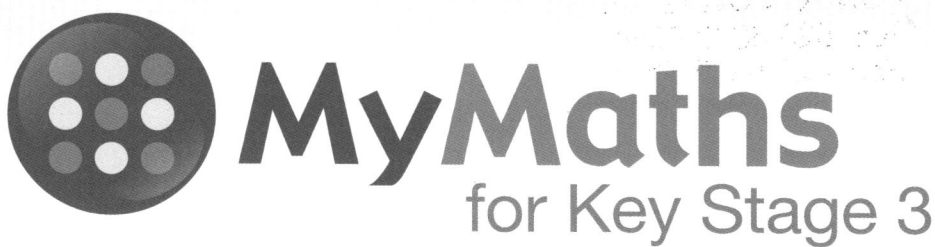

MyMaths for Key Stage 3

3B

Powered by MyMaths.co.uk

OXFORD UNIVERSITY PRESS

OXFORD
UNIVERSITY PRESS

Great Clarendon Street, Oxford OX2 6DP

Oxford University Press is a department of the University of Oxford.
It furthers the University's objective of excellence in research, scholarship,
and education by publishing worldwide in

Oxford   New York
Auckland   Cape Town   Dar es Salaam   Hong Kong   Karachi
Kuala Lumpur   Madrid   Melbourne   Mexico City   Nairobi
New Delhi   Shanghai   Taipei   Toronto

With offices in

Argentina   Austria   Brazil   Chile   Czech Republic   France   Greece
Guatemala   Hungary   Italy   Japan   South Korea   Poland   Portugal
Singapore   Switzerland   Thailand   Turkey   Ukraine   Vietnam

Oxford is a registered trade mark of Oxford University Press
in the UK and in certain other countries

British Library Cataloguing in Publication Data

Data available

ISBN 978-0-19-830463-0
10 9 8 7 6 5 4 3 2

Printed in Great Britain

Contents

1	**Number** Whole numbers and decimals	1
2	**Geometry and measures** Measure and area	5
3	**Algebra** Expressions and formulae	11
4	**Number** Fractions decimals and percentages	16
5	**Geometry and measures** Angles and 2D shapes	23
6	**Algebra** Graphs	28
7	**Number** Decimal calculations	37
8	**Statistics and probability** Statistics	42
9	**Geometry and measures** Transformations and scale	53
10	**Algebra** Equations	58
11	**Number** Powers and roots	63
12	**Geometry and measures** Constructions and Pythagoras	68
13	**Algebra** Sequences	73
14	**Geometry and measures** 3D shapes	78
15	**Ratio and proportion** Ratio and proportion	83
16	**Statistics and probability** Probability	90
	Glossary	97

Example

Work out **a** 1.9×10^3

b $8^9 \div 8^2$, leaving your answer as a power of 8.

a $1.9 \times 10^3 = 1.9 \times 1000 = 1900$

(The decimal point is moved 3 places to the right.)

b $8^9 \div 8^2 = 8^7$ (The indices are subtracted when dividing.)

1 Calculate these.
a	**i**	5×100	**ii**	3.5×100	**iii**	0.25×100
b	**i**	7×1000	**ii**	4.2×1000	**iii**	0.39×1000
c	**i**	8×10000	**ii**	6.3×10000	**iii**	0.41×10000
d	**i**	12×10000	**ii**	13.5×10000	**iii**	1.47×10000

2 Calculate these.
a	**i**	$36 \div 100$	**ii**	$5.9 \div 100$	**iii**	$0.93 \div 100$
b	**i**	$27 \div 1000$	**ii**	$3.6 \div 1000$	**iii**	$0.57 \div 1000$

3 Calculate these.
a	**i**	8×0.1	**ii**	0.8×0.1	**iii**	800×0.1
b	**i**	$9 \div 0.01$	**ii**	$0.09 \div 0.01$	**iii**	$900 \div 0.01$

4 Find the missing number.

a **i** $5.6 \times \square = 56$ **ii** $5.6 \times \square = 5600$ **iii** $5.6 \times \square = 56000$

b **i** $8.2 \div \square = 0.82$ **ii** $8.2 \div \square = 0.082$ **iii** $8.2 \div \square = 0.00082$

c **i** $14.6 \div \square = 0.146$ **ii** $14.6 \div \square = 0.00146$ **iii** $14.6 \div \square = 0.0000146$

5 Work out each of these.
a	3.2×10^2	**b**	5.1×10^5	**c**	7.3×10^6
d	1.25×10^3	**e**	4.16×10^5	**f**	6.49×10^4

6 Evaluate these, leaving your answer as a power of the number.
a	$10^4 \times 10^3$	**b**	$6^5 \times 6^2$	**c**	$5^3 \times 5$
d	$9^7 \div 9^5$	**e**	$4^8 \div 4^2$	**f**	$7^5 \div 7$

Example

Use a calculator to evaluate $83 \div 32$. Give your answer correct to
a two decimal places **b** one decimal place.

---

a $83 \div 32 = 2.59\underline{3}75$, therefore the answer to two decimal places is 2.59, because the third decimal digit (underlined) is ignored as it is less than 5.

b $83 \div 32 = 2.5\underline{9}375$, therefore the answer to one decimal place is 2.6, because the second decimal digit (underlined) is greater than 5, so the first decimal digit is increased by 1.

1 Correct these numbers to
  i the nearest 10 **ii** the nearest 100 **iii** the nearest 1000.
  a 5678 **b** 6429 **c** 2863 **d** 7521 **e** 4396

2 Correct these numbers to
  i 2 decimal places **ii** 1 decimal place **iii** the nearest whole number.
  a 4.369 **b** 6.873 **c** 9.405 **d** 3.968 **e** 7.997

3 Use a calculator to work out these divisions. Give your answers to two decimal places.
  a $71 \div 125$ **b** $57 \div 125$ **c** $39 \div 125$ **d** $13 \div 16$
  e $11 \div 32$ **f** $23 \div 32$ **g** $19 \div 80$ **h** $7 \div 40$

4 Leith Hill in Surrey is 294 m high and the mountain Mam Sodhail in the north of Scotland is 1177 m high. By rounding these figures to the nearest hundred, find approximately how many times higher Mam Sodhail is than Leith Hill.

5 For these quantities state the limits between which they lie.
  a The distance from London to Birmingham is 175 km. (Correct to the nearest kilometre.)
  b Zeke has a mass of 71.8 kg. (Correct to one decimal place.)
  c Tnisha's street is 920 m long. (Correct to the nearest 10 metres.)

MyMaths.co.uk

Q 1001 SEARCH

Example

Find **a** the HCF
b the LCM of 8 and 12.
Illustrate the factors of both numbers on a Venn diagram.

- -

a The factors of 8 are 1, 2, <u>4</u> and 8 and those of 12 are 1, 2, 3, <u>4</u>, 6 and 12. 4 is the highest number that is on both lists. Therefore 4 is the HCF.

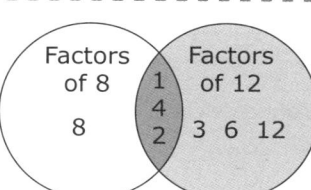

b The LCM is the lowest number that is divisible by both 8 and 12. It can be found by this formula:
LCM = product of the numbers ÷ HCF
Therefore LCM $= \dfrac{8 \times 12}{4} = 96 \div 4 = 24$

1 Find all the factor pairs for these numbers. Then list all the factors for each number.
  a 18   **b** 21   **c** 28   **d** 32

2 Find the HCF of these pairs of numbers.
  a 16 and 30 **b** 20 and 32 **c** 30 and 36 **d** 40 and 48

3 Find the LCM of the pairs of numbers in question **2**.

4 Express each of these numbers as a product of its prime factors.
  a 330   **b** 462   **c** 858   **d** 105   **e** 273
  f 195   **g** 260   **h** 204   **i** 308   **j** 495

5 i Draw a Venn diagram to show the factors of these pairs of numbers.
  a 210 and 120   **b** 240 and 135   **c** 180 and 225
  d 108 and 144   **e** 78 and 130   **f** 84 and 126

  ii Use your Venn diagrams to find the HCF and the LCM for each pair of numbers in part **i**.

MyMaths.co.uk
🔍 1032   **SEARCH**
Number Whole numbers and decimals   **3**

Estimate the value of 35.6 × 8.7 by approximating the numbers to 1 significant figure.

- -

35.6 × 8.7 becomes 40 × 9 when the numbers are approximated to one significant figure; 40 × 9 = 360

1 Round each of these numbers to one significant figure.

a	2837	**b**	4356	**c**	524	**d**	679	**e**	69
f	34	**g**	75	**h**	6.1	**i**	3.6	**j**	0.323
k	0.573	**l**	0.074	**m**	0.089	**n**	0.973	**o**	96

2 Estimate the value of these by approximating the numbers to one significant figure.

a	280 × 43	**b**	36.5 × 12.7	**c**	3.87 × 2.69	**d**	0.56 × 0.32
e	345 ÷ 61.1	**f**	693.2 ÷ 7.35	**g**	56.3 ÷ 0.52	**h**	98.2 ÷ 0.044

3 Find estimates for these by approximating the numbers to one significant figure.

a The area of the field.

b The mass of 28 sheep on a wagon if their mean mass is 35.5 kg.

129 m

255 m

c The mass of 36 cars on a ferry if their mean mass is 2.15 tonnes.

d The side length of the car park.

78 m Area = 10000 m²

e The average speed of a train in km/h if it runs 838 km from London to Aberdeen in 7.5 hours.

f Joanna's average speed in metres per second if she runs 800 m in 2 minutes 53 seconds.

MyMaths.co.uk

Q 1005 **SEARCH**

2a   Measures 1

Change   **a** 500 m² to hectares   **b** 0.06 litres to centilitres.

a   500 m² = 500 ÷ 10 000 = 0.05 ha
b   0.06 litres = 0.06 × 100 cl = 6 cl

1   Copy and complete the tables.

a

mm	cm	m	km
	500		
	7000		
		800	
		2000	
600			
			25

b

cm³ (or ml)	Litres	m³
	5000	
900 000		
		0.06
8000		
	0.3	
		0.00007

2   A table top measures 120 cm by 90 cm.
Find its area in
   a   cm²   **b**   m².

120 cm

90 cm

3   A water tank measures 100 cm by 60 cm by 50 cm.
   a   Find its volume in
      i   cm³   **ii**   m³.
   b   How many litres of water are
      required to fill it?
   c   If 1 litre of water has a mass of 1 kg,
      what is the mass of the water that fills
      the tank in tonnes?

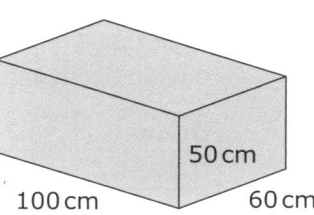

50 cm

100 cm   60 cm

Example

Using the conversion that 1 litre equals 1.75 pints, change 5 litres to pints. Give your answer correct to the nearest pint.

5 litres = 5 × 1.75 pints = 8.75 or 9 pints correct to the nearest pint. (0.75 is greater than 0.5.)

1 Copy and complete the table. Use the conversions that 1 inch equals 2.5 cm and that 1 yard equals 0.9 m.

Inches	72				126			
Centimetres		270					1350	
Feet			13.5			12		
Yards				2.5				
Metres								45

2 a Given that 1 litre equals 1.75 pints, copy and complete the table.

Litres	12	20	8			
Pints				28	17.5	10.5

b Given that 1 gallon equals 4.5 litres, copy and complete the table.

Gallons	4	7	9			
Pints				36	54	11.25

3 a Given that 1 kg equals 2.2 pounds (lb), copy and complete the table.

kg	20	2.5	4.5			
Pounds (lb)				33	16.5	38.5

b Given that 1 ounce equals 30 g, copy and complete the table.

Grams	2400	1440	3600						
Ounces				64	96	56			
Pounds (lb)							7	4.5	10.5

MyMaths.co.uk

Q 1191   SEARCH

Example

Find the height (h) of the triangle illustrated.

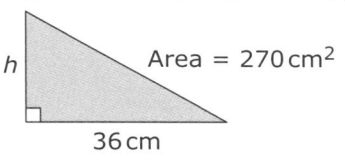

Area = 270 cm²

The area = $\frac{1}{2}bh$, therefore $\frac{1}{2} \times 36h = 270$, so $18h = 270$,

hence $h = 270 \div 18 = 15$ cm

1 Find the area of each of these by any appropriate method.

a 5 cm   8 cm

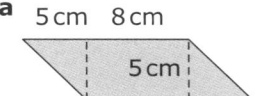

5 cm

5 cm

b 7 cm   9 cm

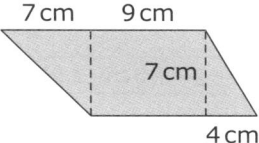

7 cm

4 cm

c

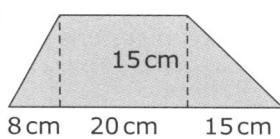

15 cm

8 cm   20 cm   15 cm

d

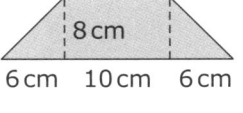

8 cm

6 cm   10 cm   6 cm

e 3 cm

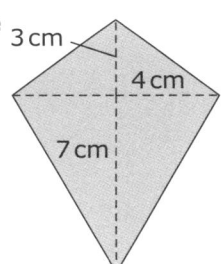

4 cm

7 cm

f

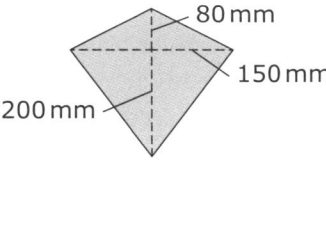

80 mm

150 mm

200 mm

2 a Find the area of the table top illustrated.

150 cm

90 cm

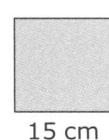

15 cm

30 cm

18 cm

b Find the number of square tiles of side length 15 cm that would fit on the table top.

c Find the number of table mats of dimensions 30 cm by 18 cm that would fit on the table top.

MyMaths.co.uk

Q 1108, 1128 **SEARCH**

Example

A button has a radius of 1.5 cm.
Find
a   its diameter
b   its circumference.

1.5 cm

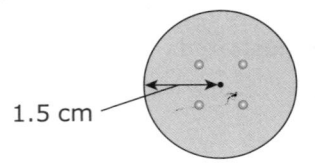

a   Diameter = 2 × radius = 2 × 1.5 = 3 cm
b   Circumference = π × diameter ≐ 3.14 × 3 = 9.42 cm

1   Copy and complete the table about
    the dimensions of circles.

Radius			40 cm	1.25 m	
Diameter	75 cm	210 mm			
Circumference					47.1 cm

Radius

Diameter

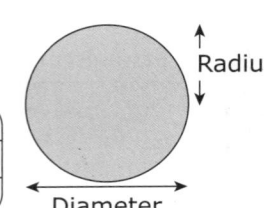

2   The Earth is 150 million kilometres
    from the Sun.
    Find the circumference of its orbit.

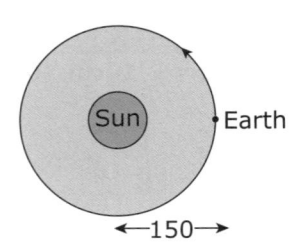

Sun   •Earth

←—150—→
million kilometres

3   A large marble has a radius of 1 cm.
    a   Find i   its diameter ii   its circumference.
    b   How many revolutions did it make when
        Jacob flicked it and it rolled for a distance
        of 157 cm?

4   Find the perimeter of
    Marlon's track for his
    toy cars.

111.6 cm

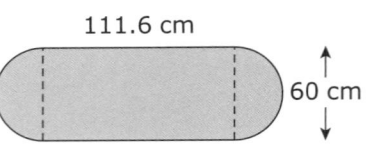

60 cm

5   Find the perimeter of this door.

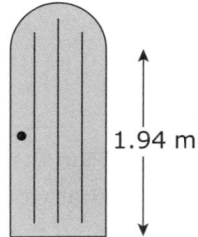

1.94 m

←1 m→

⊞ **MyMaths**.co.uk

Q 1088   **SEARCH**

2e Area of a circle

Find the surface area of the button illustrated in the worked example for Homework 2d.

Area = π × radius squared = 3.14 × 1.5 × 1.5 = 7.065 cm²

1 Find the area of each of these circles.

   a $r = 5$ cm   **b** $r = 8$ cm

   c $r = 25$ cm   **d** $r = 200$ mm

   e $r = 0.1$ m   **f** $r = 0.15$ m

2 A flat pan lid has a radius of 15 cm. Find its area.

3 a Find

   i the area of the circle

   ii the area of the square

   iii the area of the shaded part.

b Find

   i the area of the circle

   ii the area of the rectangle

   iii the area of the shaded part.

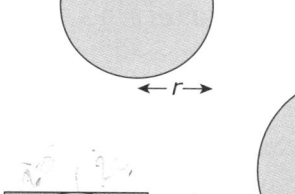

20 cm

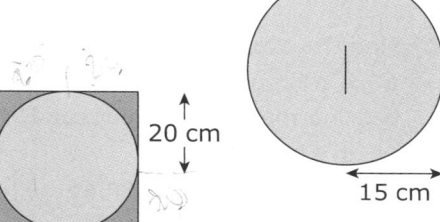

15 cm

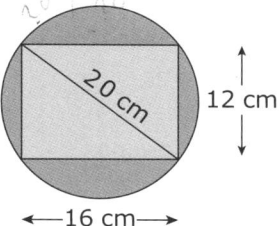

20 cm

12 cm

←——16 cm——→

4 The metal sheet illustrated is melted down and recast into discs of the same thickness with a diameter of 5 cm. Find

   i the area of the metal sheet

   ii the area of one disc

   iii the number of discs cast.

15 cm

62.8 cm

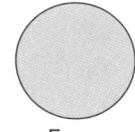

5 cm

Find the average speed, in km/h, of a motorway coach that travels 165 km from Newcastle to Leeds in 2 h 45 min.

Speed = distance ÷ time = $165 \div 2\frac{45}{60}$ (because 45 min = $\frac{45}{60}$ h)

= 60 km/h

1 Find the average speed, in metres per second, for these runners.
  a Ayo sprints 100 m in 12.5 seconds.
  b Zoey completes a 400 m race in 62.5 seconds.
  c Frank completes a 1500 m race in 4 min 10 s.

2 Find the average speed, in kilometres per hour, for these journeys.
  a A train travels 190 km from London to Bristol in 1 h 15 min.
  b Charmaine drives her car 154 km from London to Leicester in 1 h 45 min.
  c Ronnie rides his bike 45 km from London to Guildford in 2 h 15 min.

3 a A wooden block has a mass of 7.2 kg and its dimensions are 30 cm, 20 cm and 15 cm.
    Find
    i its mass in grams
    ii its volume in cm^3
    iii its density in grams per cubic centimetre
    iv its volume in m^3
    v its density in kilograms per cubic metre.
  b Find the mass of another block of the same kind of wood if its volume is
    i 5000 cm^3
    ii 0.002 m^3.

MyMaths.co.uk

Q 1121, 1246 SEARCH

3a Factors in algebra

Example

Factorise $15x^3 - 10x$.

$15x^3 - 10x = 5x(3x^2 - 2)$
5 and x are both factors of the expression.

1 Factorise these expressions.

a $2x + 6$	**b** $3y + 12$	**c** $5y + 30$
d $9t + 45$	**e** $4u - 20$	**f** $7v - 42$
g $8w - 64$	**h** $15m - 20$	**i** $16n - 30$
j $20p - 25$	**k** $12r + 28$	**l** $14s + 35$
m $9x + 3$	**n** $24y + 4$	**o** $27 - 9z$
p $32 - 8a$	**q** $40 - 5b$	**r** $14 - 35c$

2 Factorise these expressions.

a $x^2 + 9x$	**b** $y^2 + 4y$	**c** $z^2 - 7z$
d $t^2 - 8t$	**e** $8m^2 - 10m$	**f** $9n^2 + 12n$
g $u^3 + 5uv$	**h** $m^2 + 7m$	**i** $a^2 - 8ab$
j $u^3 + 5u^2 - 7u$	**k** $v^3 - 2v^2 + 9v$	**l** $w^3 - 6w^2 - 8w$

3 Factorise these expressions. In all cases there are two common factors.

a $2x^2 + 10xy$	**b** $5u^2 + 20uv$	**c** $3m^2 - 21mn$
d $7p^2 - 35pq$	**e** $8a^2 + 10ab$	**f** $9c^2 + 12cd$
g $10x^2 + 25xy$	**h** $14u^2 - 21uv$	**i** $12z^3 + 10z$
j $15u^3 + 10u$	**k** $18v^3 + 9v$	**l** $16p^3 - 12p$

4 The cuboid illustrated has a volume which is given by
$14x^2 - 24x$.

a Factorise the expression fully.

b Write down the three side dimensions of the cuboid if they are the same as the three factors of the expression.

c If $x = 2\,\text{cm}$, what can be said about the cuboid?

Example

Simplify $\frac{3a}{10} + \frac{4b}{15}$

- -

$\frac{3a}{10} + \frac{4b}{15} = \frac{9a}{30} + \frac{8b}{30} = \frac{(9a+8b)}{30}$

1 Simplify these fractions as much as possible.

    a $\frac{6x}{9}$ **b** $\frac{4y}{10}$ **c** $\frac{5z}{20}$

    d $\frac{xy}{5y}$ **e** $\frac{uv}{9u}$ **f** $\frac{9ab}{a^2}$

    g $\frac{7pq}{q^2}$ **h** $\frac{3m}{5mn}$ **i** $\frac{8x}{16xy}$

2 Add or subtract these fractions.

    a $\frac{4x}{5} + \frac{3x}{5}$ **b** $\frac{2y}{9} + \frac{5y}{9}$ **c** $\frac{5a}{7} + \frac{a}{7}$ **d** $\frac{11x}{15} - \frac{7x}{15}$

    e $\frac{8y}{11} - \frac{3y}{11}$ **f** $\frac{3b}{8} + \frac{c}{8}$ **g** $\frac{4p}{9} + \frac{q}{9}$ **h** $\frac{8u}{15} - \frac{v}{15}$

    i $\frac{9p}{13} - \frac{4p}{13}$ **j** $\frac{5q}{9} - \frac{4q}{9}$

3 **a** Find an expression for

      i   how far Afiya lives from the school

      ii how far Eshe lives from the school.

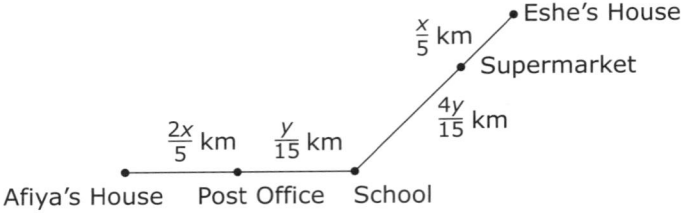

    b Inez says that as they both live the same distance from the school
       it follows that $x = y = 2$.
       Is Inez right? Explain your answer.

MyMaths.co.uk

Q 1178   **SEARCH**

3c Formulae in context

Example

The volume (V) of a cuboid is given by $V = lwh$.

a Find V if $l = 4$, $w = 3$ and $h = 2.5$.

b Find h if $V = 63$, $l = 7$ and $w = 6$.

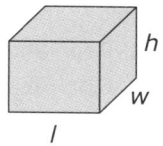

a $V = 4 \times 3 \times 2.5 = 30$

b $63 = 7 \times 6 \times h$ or $63 = 42h$, so $h = 63 \div 42 = 1.5$

---

1 The stretched length of a spring (L) is given by $L = l + st$, where l is the natural length of the spring, s is the spring constant and t is the mass hanging.

Find L if $l = 20$, $s = 0.1$ and t is equal to

  a 50 **b** 200 **c** 125

2 The perimeter (P) of the isosceles trapezium illustrated is given by $P = 4l + 2m$.

Find P if

  a $l = 50$ and $m = 70$

  b $l = 35$ and $m = 50$

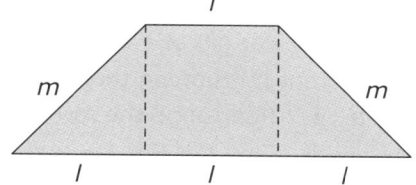

3 The area (A) of the isosceles trapezium in question **2** is given by $A = 2l^2$.

Find A if l is equal to **a** 15 **b** 4.5 **c** 0.25

4 The area (A) of a kite is given by the formula $A = \frac{1}{2}xy$ where x and y are the diagonal lengths.

Find A for the two kites illustrated.

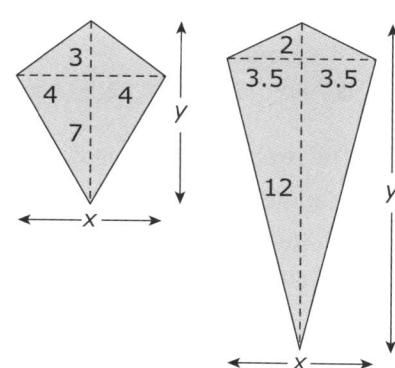

5 For the spring in question **1** find

  a l if $L = 35$, $t = 150$ and $s = 0.12$

  b t if $L = 70$, $l = 40$ and $s = 0.15$

  c s if $L = 90$, $l = 75$ and $t = 60$

Example

Look at the cuboid in the worked example for Homework 3c.

a Rearrange the volume formula so as to make l the subject.

b Find l if $V = 22.5$, $w = 3$ and $h = 2.5$.

- -

a $V = lwh$, therefore $l = \dfrac{V}{wh}$. (Both sides have been divided by wh.)

b $l = \dfrac{V}{wh} = 22.5 \div (3 \times 2.5) = 22.5 \div 7.5 = 3$

1 Speed (s), distance (d) and time (t) are related by the formula $s = \dfrac{d}{t}$.

  a **i** Rearrange the formula to make d the subject.

     ii Find d if $s = 56$ and $t = 4$.

  b **i** Rearrange the formula to make t the subject.

     ii Find t if $d = 315$ and $s = 63$.

2 The power (P) of a lamp, the voltage (V) applied to it and the current (I) running through it are related by the formula $P = VI$.

  a **i** Rearrange the formula to make V the subject.

     ii Find V if $P = 1.8$ and $I = 0.12$.

  b **i** Rearrange the formula to make I the subject.

     ii Find I if $P = 1.5$ and $V = 12$.

3 The volume (V) of the wedge illustrated is given by $V = \dfrac{1}{2}abc$.

  a Rearrange the formula to make a the subject.

  b Find a if $V = 112.5$, $b = 5$ and $c = 6$.

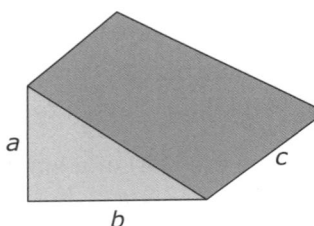

4 The area (A) of the isosceles right-angled triangle illustrated is given by $A = \dfrac{1}{4}b^2$.

  a Rearrange the formula to make b the subject.

  b Find b if $A = 6.25$.

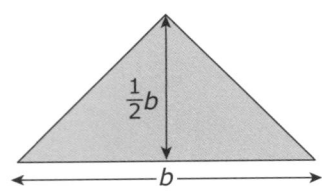

Example

An examination board charges an entry fee of £20 plus £15 per subject taken.

a Find a formula for the cost (£C) in terms of n the number of subjects taken.

b Find the value of C for n equal to 0, 2, 4, 6 and 8.

- -

a $C = 20 + 15n$

b

n	0	2	4	6	8
C (£)	20	50	80	110	140

1 **a** A running track has a 100 m 'run on' and a 300 m circuit. If a runner starts from A and goes on to complete n laps of the circuit, find a formula for d the distance covered in terms of n.

Perimeter 300 m

A
← 100 m →

b Find the value of d for n equal to 1, 2, 3, 4 and 5 and plot the details on a graph.

2 **a** The figure shown for a regular hexagon is true to a very close approximation. Assuming the figure to be correct, show that the area (A cm²) is given by $A = \dfrac{21l^2}{8}$.

b Find the value of A for l equal to 2, 4, 6 and 8 cm and plot the details on a graph.

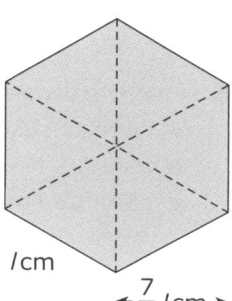

l cm

$\leftarrow \dfrac{7}{8} l$ cm $\rightarrow$

4a Adding and subtracting fractions

Example

Work out **a** $\dfrac{9}{10} + \dfrac{4}{15}$

b $1\dfrac{1}{3} - \dfrac{4}{5}$

a $\dfrac{9}{10} + \dfrac{4}{15} = \dfrac{27}{30} + \dfrac{8}{30} = \dfrac{35}{30} = \dfrac{7}{6} = 1\dfrac{1}{6}$

b $1\dfrac{1}{3} - \dfrac{4}{5} = \dfrac{4}{3} - \dfrac{4}{5} = \dfrac{20}{15} - \dfrac{12}{15} = \dfrac{8}{15}$

1 Calculate these. Give each answer in its simplest form.

a $\dfrac{2}{5} + \dfrac{3}{10}$ **b** $\dfrac{3}{8} + \dfrac{5}{16}$ **c** $\dfrac{2}{5} + \dfrac{7}{20}$ **d** $\dfrac{1}{2} + \dfrac{3}{10}$

e $\dfrac{3}{4} - \dfrac{1}{12}$ **f** $\dfrac{5}{6} - \dfrac{1}{12}$ **g** $\dfrac{2}{3} - \dfrac{5}{12}$ **h** $\dfrac{3}{4} - \dfrac{1}{20}$

2 Calculate these, giving your answer as a mixed number in its simplest form.

a $\dfrac{4}{5} + \dfrac{3}{10}$ **b** $\dfrac{3}{4} + \dfrac{5}{8}$ **c** $\dfrac{3}{5} + \dfrac{13}{20}$ **d** $\dfrac{5}{6} + \dfrac{11}{12}$

3 Calculate these. Give each answer in its simplest form and as a mixed number where appropriate.

a $\dfrac{7}{10} + \dfrac{2}{15}$ **b** $\dfrac{1}{6} + \dfrac{8}{15}$ **c** $\dfrac{5}{12} + \dfrac{1}{20}$ **d** $\dfrac{7}{12} + \dfrac{7}{10}$

e $\dfrac{9}{10} - \dfrac{5}{6}$ **f** $\dfrac{9}{10} - \dfrac{1}{15}$ **g** $\dfrac{7}{8} - \dfrac{5}{6}$ **h** $\dfrac{14}{15} + \dfrac{3}{20}$

4 Calculate these. Give each answer as a mixed number in its simplest form.

a $1\dfrac{1}{2} + \dfrac{4}{5}$ **b** $1\dfrac{3}{4} + \dfrac{5}{6}$ **c** $1\dfrac{3}{10} + 1\dfrac{1}{4}$ **d** $1\dfrac{1}{3} + 1\dfrac{2}{5}$

e $2\dfrac{2}{5} - 1\dfrac{1}{5}$ **f** $2\dfrac{3}{4} - 1\dfrac{3}{10}$ **g** $2\dfrac{1}{6} - \dfrac{9}{10}$ **h** $2\dfrac{1}{10} - \dfrac{4}{15}$

5 Find the distance from
a Ayo's house to Jisanne's house
b Shani's house to Jisanne's house
c Shani's house to Candace's house.

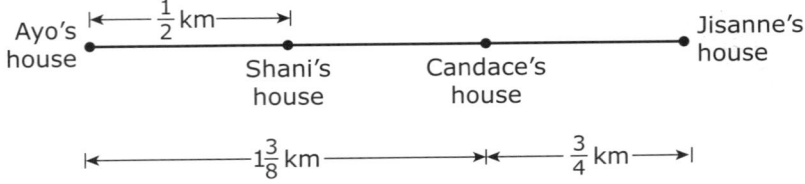

Example

a Calculate $\frac{5}{6} \times \frac{3}{8}$

b A car with catalogue price £10000 is advertised in a sale at $\frac{3}{4}$ of this price. What is its sale price?

---

a $\frac{5}{6} \times \frac{3}{8} = \frac{15}{48} = \frac{5}{16}$

b The sale price is $\frac{3}{4} \times 10000 = \frac{30000}{4} = £7500$.

1 Calculate these. Give your answer in its simplest form and as a mixed number where appropriate.

a $3 \times \frac{2}{7}$ b $5 \times \frac{3}{16}$ c $2 \times \frac{4}{9}$ d $4 \times \frac{3}{20}$

e $3 \times \frac{5}{24}$ f $6 \times \frac{5}{42}$ g $4 \times \frac{3}{7}$ h $7 \times \frac{3}{20}$

2 Calculate these. Give your answer in its simplest form and as a mixed number where appropriate.

a $\frac{3}{5}$ of 12 cm   b $\frac{2}{3}$ of 16 kg   c $\frac{3}{4}$ of 25 m

d $\frac{5}{6}$ of 30 min   e $\frac{4}{5}$ of 35 grams   f $\frac{3}{4}$ of 28 km

3 Calculate these. Give your answers as mixed numbers where appropriate.

a $1\frac{1}{2} \times 7$ m   b $1\frac{1}{3} \times 5$ km   c $1\frac{1}{4} \times 3$ kg   d $1\frac{3}{4} \times 5$ ml

e $1\frac{1}{4} \times 8$ min   f $1\frac{1}{3} \times 9$ s   g $1\frac{2}{5} \times 15$ g   h $1\frac{2}{3} \times 18$ hours

4 a Barbara lives $\frac{5}{6}$ km from her school and one day she runs for $\frac{3}{4}$ of the way. How far does she run?

b Josiah wants to buy a bike which normally costs £250, but he buys it for $\frac{4}{5}$ of that price in a sale. How much did he pay?

c Zeke buys a car for £12000 but later he sells it for $\frac{5}{8}$ of this price. How much did he sell it for?

MyMaths.co.uk

Q 1047   **SEARCH**

4c Dividing by fractions

Calculate **a** $3 \div \frac{12}{17}$ **b** $\frac{5}{9} \div \frac{15}{16}$

a $3 \div \frac{12}{17} = 3 \times \frac{17}{12} = \frac{51}{12} = \frac{17}{4} = 4\frac{1}{4}$

b $\frac{5}{9} \div \frac{15}{16} = \frac{5}{9} \times \frac{16}{15} = \frac{80}{135} = \frac{16}{27}$

---

1 Calculate these. Express your answer as a mixed number.

   a $7 \div \frac{2}{3}$ **b** $4 \div \frac{5}{9}$ **c** $7 \div \frac{3}{4}$ **d** $5 \div \frac{7}{10}$

   e $5 \div \frac{3}{4}$ **f** $3 \div \frac{5}{9}$ **g** $6 \div \frac{5}{8}$ **h** $3 \div \frac{4}{9}$

2 Calculate these.

   a $8 \div \frac{4}{5}$ **b** $12 \div \frac{2}{3}$ **c** $8 \div \frac{2}{5}$ **d** $15 \div \frac{3}{5}$

   e $12 \div \frac{4}{5}$ **f** $9 \div \frac{3}{4}$ **g** $12 \div \frac{3}{7}$ **h** $9 \div \frac{3}{11}$

3 Calculate these. Express your answer as a mixed number.

   a $3 \div \frac{6}{7}$ **b** $4 \div \frac{8}{11}$ **c** $5 \div \frac{10}{13}$ **d** $3 \div \frac{9}{10}$

   e $5 \div \frac{15}{16}$ **f** $4 \div \frac{16}{23}$ **g** $4 \div \frac{12}{17}$ **h** $3 \div \frac{12}{19}$

4 Calculate these. Give your answer in its simplest form
and express it as a mixed number where appropriate.

   a $\frac{2}{7} \div \frac{3}{8}$ **b** $\frac{3}{8} \div \frac{4}{5}$ **c** $\frac{4}{9} \div \frac{5}{7}$ **d** $\frac{4}{5} \div \frac{8}{9}$

   e $\frac{4}{7} \div \frac{12}{13}$ **f** $\frac{3}{5} \div \frac{9}{11}$ **g** $\frac{5}{8} \div \frac{20}{21}$ **h** $\frac{3}{5} \div \frac{4}{7}$

5 **a** A hardware store has a drum containing 15 litres of
paraffin. If it is used to fill cans with capacity $\frac{5}{6}$ of a
litre, how many cans will it fill?

   b Jake requires some pieces of string of length $\frac{3}{4}$ m.
If he has a string ball with a label marked '12 m', how
many pieces can he cut?

MyMaths.co.uk

Q 1040 **SEARCH**

Example

a Change the decimal 0.725 to a fraction in its simplest form.

b Change the fraction $\frac{10}{11}$ to a decimal. Give your answer to three decimal places.

---

a $0.725 = \frac{725}{1000} = \frac{145}{200} = \frac{29}{40}$

b $\frac{10}{11} = 10 \div 11 = 0.909\underline{0}90... = 0.909$ to three decimal places.

1 Change these decimals to fractions in their simplest form.

a 0.78	**b** 0.46	**c** 0.14	**d** 0.64	**e** 0.96
f 0.08	**g** 0.225	**h** 0.075	**i** 0.015	**j** 1.275

2 Change these fractions to decimals.

a $\frac{5}{8}$	**b** $\frac{7}{8}$	**c** $\frac{9}{20}$	**d** $\frac{13}{20}$	**e** $\frac{3}{4}$
f $\frac{3}{5}$	**g** $\frac{2}{5}$	**h** $\frac{1}{5}$	**i** $\frac{1}{20}$	**j** $\frac{1}{50}$
k $\frac{1}{40}$	**l** $1\frac{3}{20}$	**m** $2\frac{1}{8}$	**n** $3\frac{3}{5}$	**o** $5\frac{4}{5}$

3 **a** Change these fractions to decimals (correct to three decimal places) and arrange them in order of size, starting with the smallest.

$\frac{3}{8}, \frac{4}{9}, \frac{2}{7}, \frac{5}{11}, \frac{9}{20}$ and $\frac{11}{25}$

b Change these fractions to decimals (correct to three decimal places) and arrange them in order of size, starting with the largest.

$\frac{5}{8}, \frac{7}{11}, \frac{4}{7}, \frac{13}{20}, \frac{2}{3}$ and $\frac{16}{25}$

4 A table top measures $1\frac{4}{25}$ m by $\frac{19}{20}$ m. Marcus measures the two dimensions with a tape and says that they are 116 cm and 95 cm. Has Marcus measured them correctly?

$1\frac{4}{25}$ m

$\frac{19}{20}$ m

Example

In September 2008, Willow Lane School had 520 pupils, but by January 2009 it had 5% more.
How many pupils did the school have by January 2009?

The extra number of pupils was 5% of 520 = $\frac{5}{100} \times 520 = 26$. Therefore by January 2009 the number of pupils was 520 + 26 = 546.

1 Calculate these.

   a 15% of £60   **b** 12% of 150 cm   **c** 30% of 50 kg

   d 45% of 80 m   **e** 18% of 25 m   **f** 24% of 105 kg

   g 25% of £90   **h** 36% of £45

2 Work out these increases.

   a £60 up 25%   **b** £75 up 12%   **c** 84 m up 15%

   d 65 m up 16%   **e** 45 kg up 24%   **f** £25 up 30%

3 Work out these decreases.

   a £80 down 15%   **b** £120 down 35%   **c** 50 m down 16%

   d 54 m down 25%   **e** 48 kg down 5%   **f** £45 down 8%

4 A small school has only five classes. Copy and complete the table about the number of pupils in each class.

Class	Number of pupils
1	20
2	20% more than Class 1
3	25% more than Class 2
4	20% more than Class 3
5	25% more than Class 4

MyMaths.co.uk   1060   SEARCH

Example

a Ela earns £25 000 per year and she gets a pay rise of 4%. Ayako earns £24 000 and she gets a pay rise of 5%. Whose pay rise is the bigger and by how much?

b Steve earns £22 500 per year and he gets a pay rise of £810. Express his pay rise as a percentage.

- -

a Ela's pay rise = $\frac{4}{100} \times 25000 = £1000$

Ayako's pay rise = $\frac{5}{100} \times 24000 = £1200$

Therefore Ayako's pay rise is the bigger by £200.

b Steve's pay rise as a percentage = $\frac{810}{22500} \times 100\% = 3.6\%$

1 Last year the Johnson family earned £32 000 and spent £2880 on a car. The Gray family, however, earned £36 000 and spent £3150 on a car. Express the proportion of their income that each family spent on a car as
a a fraction **b** a decimal **c** a percentage.
Which family spent the bigger proportion of their income on a car and by how many percent?

2 Anwar invests £1500 in his bank and receives £90 interest after one year. Rapinda invests £1680 in her bank and receives £105 interest after one year. Whose bank pays the higher interest rate when expressed as a percentage and by how many percent?

3 a Zeke was 150 cm tall one year ago, but he has since grown 2%.
Find **i** how much he has grown **ii** how tall he is now.

b Zodia had a mass of 75 kg a month ago, but she has been on a slimming diet and has lost 6% of this mass.
Find **i** the mass she has lost **ii** what her mass is now.

4 On a market stall a 420 g packet of currants costs 84p. Another packet that contains 20% more currants costs 90p. Which is the better buy and by how many grams per penny?

Example

Jim buys a bike for £200, but it depreciates by 10% per annum. How much is it worth after 2 years?

After 1 year its value is 90% of £200 or $\frac{90}{100} \times$ £200, which is £180.

Therefore after 2 years its value is 90% of £180 or $\frac{90}{100} \times$ £180, which is £162.

1 Calculate these percentage changes.

a The value of a house after 5 years which was bought for £200 000 and its value increases by 10% per annum.

b The value of a holiday home after 3 years which was bought for £80 000 and its value increases by 5% per annum.

c The value of a car after 4 years which was bought for £20 000 and its value decreases by 10% per annum.

d The value of a van after 4 years which was bought for £15 000 and its value decreases by 20% per annum.

2 The table shows details of price reductions at a sale. Copy and complete the table.

Article	Original price	Percentage reduction	Actual reduction	Sale price
Television	£360	20%		
Set of chairs	£70	15%		
Table	£120			£90
Cooker	£320			£280
Cabinet		12%		£132
Microwave oven		16%		£75.60
Lawn mower	£350		£35	
Bike	£220		£44	

MyMaths.co.uk

Q 1073 SEARCH

Example

Find the angles marked x, y and z on the diagram illustrated.

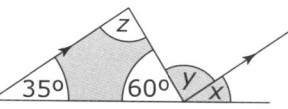

- -

$x = 35°$. (It is corresponding to the angle which is marked as 35°.)

$y = 180° - 60° - 35° = 85°$. $z = 85°$ (It is alternate to y.)

1 For each part of the question find the angles marked p, q, r, s, t, u and v.

a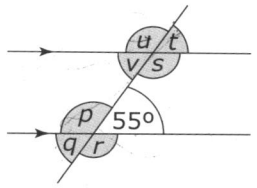

b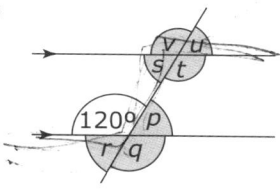

c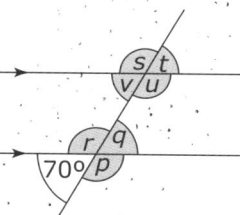

2 For each part of the question find the angles marked p, q, r, s and t.

a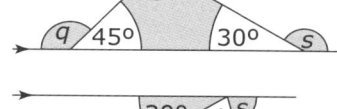

b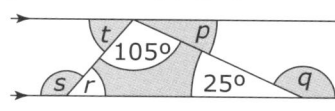

c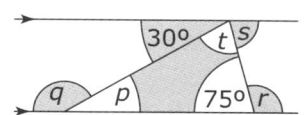

d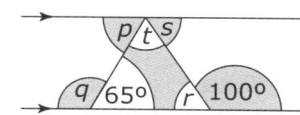

3 For each part of the question find the angles marked x, y and z.

a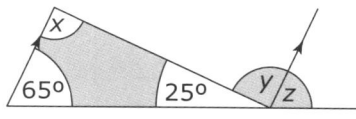

b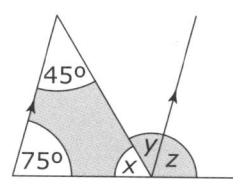

c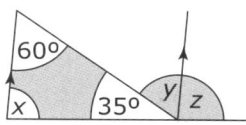

d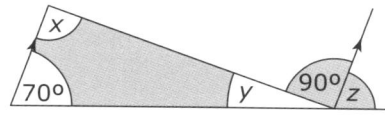

Example

a Find the fourth angle of a quadrilateral which has three angles equal to 130°, 105° and 75°.

b Could this quadrilateral be a trapezium? If so, explain why.

- -

a The fourth angle is 360° − 130° − 105° − 75° = 50°.

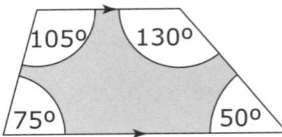   or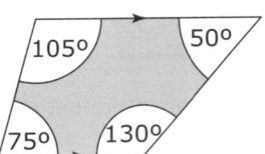

b Yes, because it has two pairs of angles which total 180°.
105° + 75° and 130° + 50° are both equal to 180°.

1 For each part of the question, find the marked angles.

a
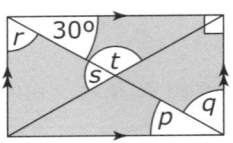

b
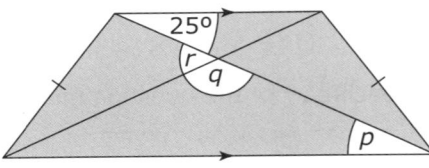

c
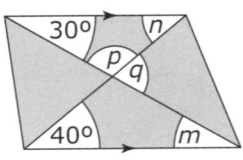

d
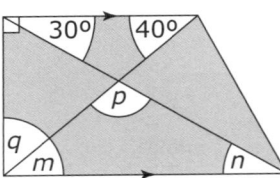

2 Four congruent triangles are illustrated.
Draw diagrams to show how the four can
be arranged to make

a a square   **b** a rectangle

c a rhombus   **d** two parallelograms

e two isosceles trapeziums   **f** an arrowhead.

Example

For the equilateral triangle illustrated find the angles *a* and *b*.

$a = 360° ÷ 3 = 120°$

$b = (180 − a)° ÷ 2 = (180 − 120)° ÷ 2$

$= 60° ÷ 2 = 30°$

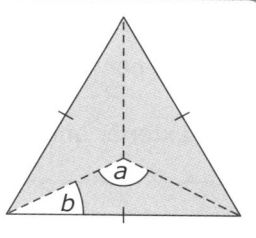

1 Make an accurate drawing of the equilateral triangle illustrated. Draw in the three axes of symmetry and measure the distance from where they cross to any vertex.

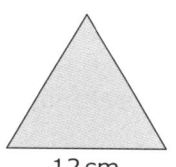

12 cm

2 Copy and complete this table for regular polygons. (The details for the regular hexagon illustrated have been done for you.)

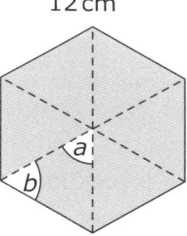

Number of sides (*n*)	Angle *a*	Angle *b*	Interior angle (2*b*)	Sum of the interior angles (2*bn*)
3				
4				
5				
6	360° ÷ 6 = 60°	(180° − 60°) ÷ 2 = 60°	2 × 60° = 120°	120° × 6 = 720°

3 Draw an accurate diagram of these two irregular polygons and draw on any symmetry axes that they have. Also state the order of rotational symmetry for both.

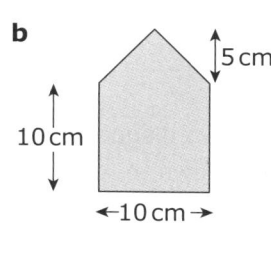

a 7 cm 8 cm 8 cm

b 5 cm 10 cm ←10 cm→

⊞ **MyMaths**.co.uk

Geometry and measures Angles and 2D shapes

Example

Find the exterior angle for a regular 7-sided polygon.

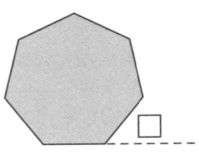

The exterior angle is $360 \div 7 = 51\frac{3}{7}$ degrees.

1 a Regular octagons together with squares can form a tessellation as can regular hexagons together with equilateral triangles. Using copies of the figures illustrated, draw the pattern for both cases.

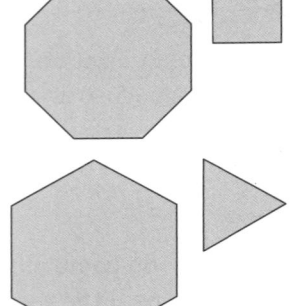

b Clever Cloggs Clarence in Class 9 found another pair of regular polygons that could form a tessellation and part of his pattern is shown. What were the two regular polygons that he found?

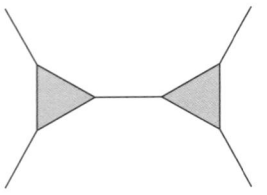

2 Three diagonals of a regular hexagon are drawn to form four triangles.
Find the sizes of angles *a*, *b*, *c*, *d* and *e*.

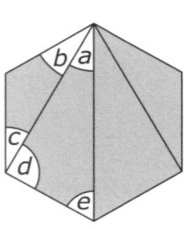

3 Two diagonals of a regular hexagon are drawn to form two triangles and a kite.
Find the four angles of the kite.

Kite

MyMaths.co.uk

Q 1320   SEARCH

Example

These isosceles triangles are congruent.

Find

a the angles p, q and r

b the side lengths AB and BC.

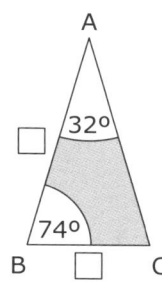
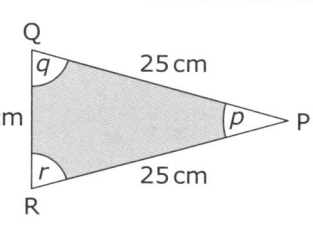

---

a $p = 32°$ (equal to the angle at A) and $q = r = 74°$ (equal to the angle at B or C)

b AB = PQ or PR = 25 cm, BC = QR = 14 cm

---

1 These triangles are congruent.

Find

a the angles m, x, y and z

b the side lengths XY, XZ and YZ.

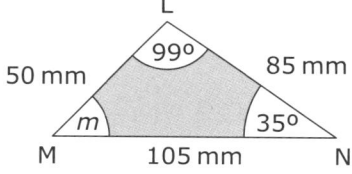
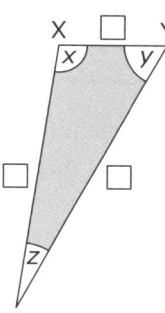

2 These trapeziums are congruent.

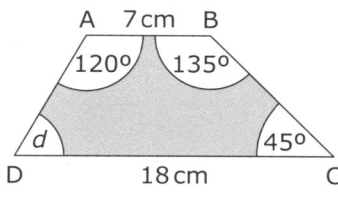
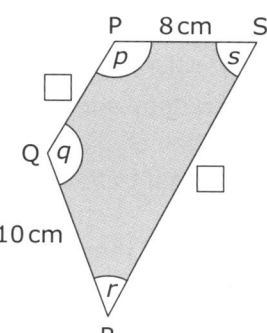

Find

a the angles d, p, q, r and s

b the side lengths AD, BC, PQ and RS.

Example

Copy and complete the graph for the equation $y = 2x + 1$.

x	0	1	2	3	4	5
y	1	3	5	7	9	11

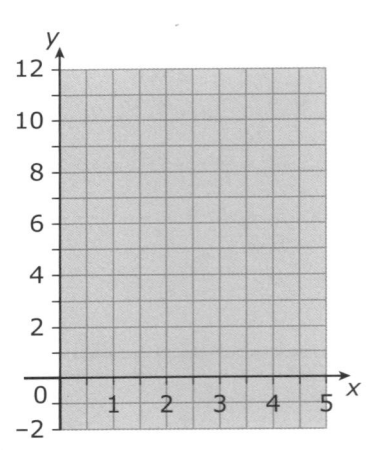

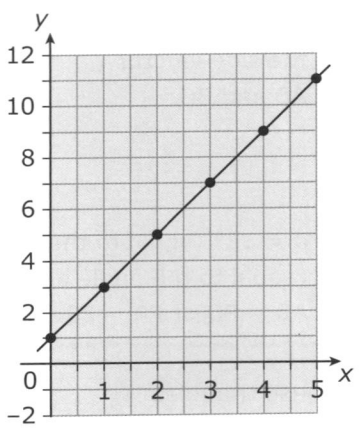

For each answer in questions **1** and **2**, a copy of the grid illustrated in the worked example is required.

1 Copy and complete the table, and then draw a graph for each of the equations.

x	0	1	2	3	4	5
y						

   a $y = x + 7$ **b** $y = 2x + 2$ **c** $y = 8 - x$

   d $y = 12 - x$ **e** $y = 12 - 2x$ **f** $y = 8 - 2x$

2 A car ferry charges £C for a crossing and C is given by $C = 3x + 1$, where x is the number of people in the car. Copy and complete the table and draw a graph of C against x. (Draw the x-axis from 0 to 5 and the C-axis from 0 to 16.)

x	0	1	3	5
C				

How many people would there be in the car if the charge was

   a £7   **b** £13?

6b   Drawing straight–line graphs

On axes labelled from 0 to 8, draw
the triangle whose sides are given
by the lines of these equations, and
find the area of the triangle.

$y = 6 - x$, $y = 2x$ and $y = 0$
(the x-axis).

- -

The area of the triangle is
$\frac{1}{2}$ × base × height = $\frac{1}{2}$ × 6 × 4
= 3 × 4 = 12 square units

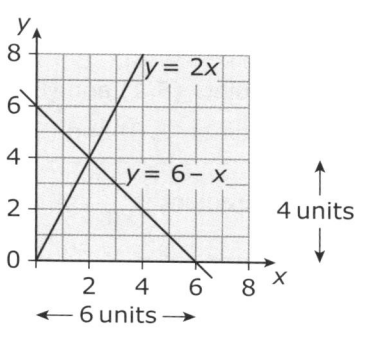

1   For each set of equations
   i   draw the four lines on axes labelled from 0 to 10
   ii shade the rectangle that you have made and find its area.
   a $x = 2$, $x = 10$, $y = 4$, $y = 10$ **b** $x = 3$, $x = 7$, $y = 1$, $y = 7$
   c $x = 1$, $x = 9$, $y = 3$, $y = 8$ **d** $x = 4$, $x = 8$, $y = 5$, $y = 10$

2   For each pair of equations
   i   copy and complete this table
          of values

x	0	2	4	6
y				

   ii draw the graphs of both lines on
          a copy of these axes
   iii find the coordinates of the point
          where the two lines intersect.
   a $y = 3x + 1$ and $y = x + 3$ **b** $y = 3x - 2$ and $y = x + 2$
   c $y = 2x - 2$ and $y = x + 3$ **d** $y = x - 2$ and $y = 6 - x$
   e $y = x + 1$ and $y = 5 - x$ **f** $y = 2x$ and $y = 6 - x$

Example

Find the gradient of the straight line that joins together

a the points $(5, 1)$ and $(8, 7)$

b the points $(1, 6)$ and $(3, 2)$.

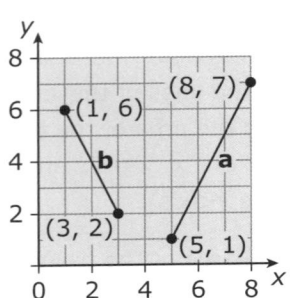

- -

a The gradient is $\frac{(7-1)}{(8-5)} = \frac{6}{3} = 2$

b The gradient is $\frac{(2-6)}{(3-1)} = \frac{-4}{2} = -2$

1 Find the gradients of these straight lines.

a 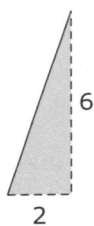 6 2

b 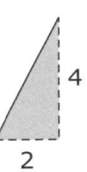 4 2

c 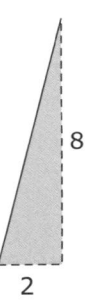 8 2

d 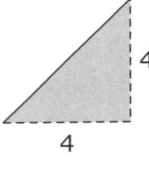 4 4

e  2 4

f −4 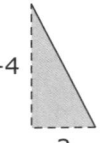 2

g −3 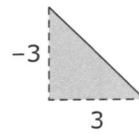 3

2 Draw and label axes from 0 to 8. Plot each pair of points, join them with a straight line and find the gradient of each straight line.

    a $(1, 2)$ and $(3, 8)$ **b** $(2, 1)$ and $(5, 7)$ **c** $(1, 0)$ and $(3, 8)$

    d $(2, 5)$ and $(4, 7)$ **e** $(0, 4)$ and $(6, 7)$ **f** $(3, 5)$ and $(6, 2)$

    g $(1, 8)$ and $(3, 2)$ **h** $(5, 4)$ and $(1, 2)$

3 Find the gradient of the straight lines with these equations.
(Do not draw any graphs.)

    a $y = 7x + 4$ **b** $y = 5x - 3$ **c** $y = \frac{1}{2}x + 10$ **d** $y = -4x + 1$

MyMaths.co.uk

Q 1312   SEARCH

Example

Plot the points $(-3, -5)$ and $(1, 3)$ on a grid and join them with a straight line. Find the y-intercept of the line.

The y-intercept is at $y = 1$, i.e. the point $(0, 1)$.

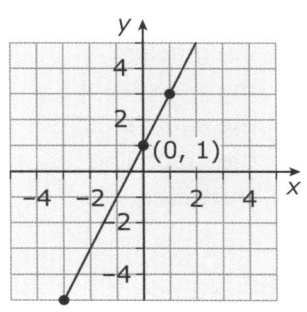

1 Draw and label axes from -5 to 5 (as in the worked example above). Plot each pair of points and join them with a straight line. Find the y-intercept for each line.

  a $(-2, -2)$ and $(1, 4)$ **b** $(-2, 1)$ and $(2, 5)$

  c $(-3, 4)$ and $(4, -3)$ **d** $(-1, 5)$ and $(3, -3)$

2 Copy and complete the table below for the four equations given.

x	0	2	4	6
$y = 2x - 1$				
$y = \frac{1}{2}x + 2$				
$y = 4 - x$				
$y = 10 - 2x$				

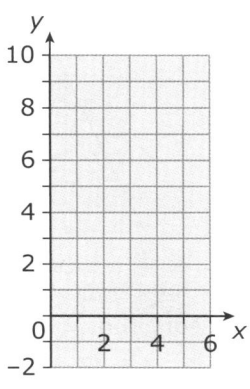

Draw their graphs on a copy of these axes and find the y-intercept for each of the lines.

3 Find the y-intercept of the straight lines with these equations. (Do not draw any graphs.)

  a $y = 4x + 7$ **b** $y = 2x + 8$ **c** $y = 3x - 4$

  d $y = \frac{1}{2}x + 5$ **e** $y = x + 9$ **f** $y = x - 6$

Example

Find

a   the gradient

b   the y-intercept

c   the equation of the graph line illustrated.

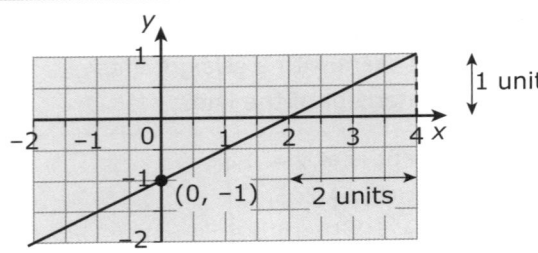

a   The gradient is $1 \div 2 = \frac{1}{2}$

b   The y-intercept is at $y = -1$

c   The equation of the line is $y = (\text{gradient})x + \text{intercept} = \frac{1}{2}x - 1$

---

1   For each of these straight lines find

   i   the gradient   **ii**   the y-intercept   **iii**   the equation.

a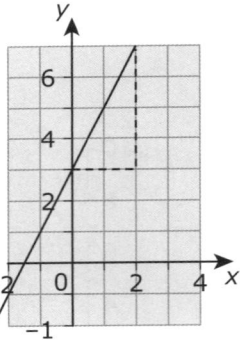

b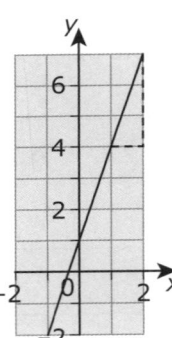

c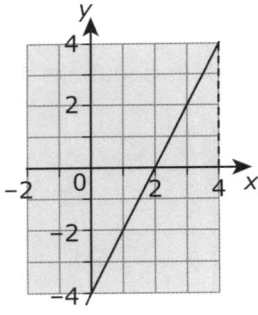

2   Find the gradient and the y-intercept of the straight lines with these equations. (Do not draw any graphs.)

   a $y = 2x + 2$ **b** $y = 4x + 3$ **c** $y = x + 4$

   d $y = 2x - 3$ **e** $y = 3x - 2$ **f** $y = \frac{1}{2}x - 4$

MyMaths.co.uk

Q 1153   **SEARCH**

Example

For the graph line of the equation $2x + 5y = 10$ find
a the intercepts on both axes   b the gradient.

- -

a At the x-intercept, $y = 0$, therefore $2x = 10$,
so $x = 5$. At the y-intercept, $x = 0$, therefore
$5y = 10$, so $y = 2$.

b $2x + 5y = 10$, therefore $5y = 10 - 2x$, so
dividing through by 5 gives $y = 2 - \frac{2}{5}x$,
hence the gradient is $-\frac{2}{5}$.

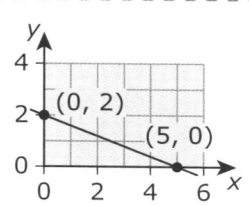

1 Find the intercepts on both axes for the
lines with these equations. Draw the lines
on a copy of the axes illustrated.
a $3x + 4y = 12$ b $3x + 2y = 6$
c $2x + 5y = 10$ d $5x + 3y = 15$
e $3x + y = 6$ f $2x + y = 4$

2 Find the intercepts on both axes for
the lines with these equations. Draw
the lines on a copy of the axes illustrated
in question **1**.
a $2x - 3y = 6$ b $4x - 5y = 20$ c $3x - 4y = 12$
d $5x - 3y = 15$ e $3x + 5y = -15$ f $4x + 3y = -12$

For questions **3** and **4** draw a pair of axes labelled from -10 to 10.
In both questions, for each equation find
    i   the intercepts on both axes   ii   the gradient.

3 a $4x + y = 8$ b $5x + y = 10$
  c $x + 3y = 9$ d $x + 2y = 6$

4 a $3x - y = 9$ b $2x - y = 8$
  c $x + 3y = -6$ d $x + 2y = -8$

Example

A spring has a natural length of 20 cm, but if a mass of
m grams (g) is hung on its end, the length (L cm) is given by
$L = 20 + \frac{1}{10} m$.

m	0	200	400
L			

a Copy and complete the
   table and draw a graph.

b Find from the graph the length of the spring when
   the hanging mass is 300 g.

a

m	0	200	400
L	20	40	60

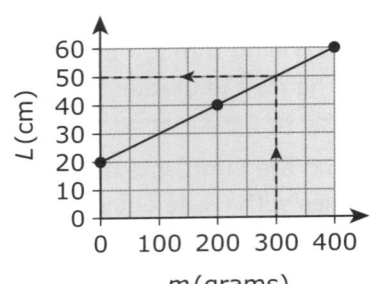

b It can be seen from the graph that for
   a hanging mass of 300 g, the length is 50 cm.

1 In a science experiment Shani is heating some water in a can. She
   records the temperature (T °C) at 30 second intervals and finds that
   the temperature is given by $T = 20 + \frac{1}{3} t$ where t is the heating time in
   seconds.

t	5	0	60	120	180	240
T	5					

   a Copy and complete the table
      and draw a graph.

   b Find from your graph the temperature after
      i 30 seconds ii 150 seconds.

   c Find from your graph the time taken for the temperature to reach
      i 50 °C   ii 90 °C.

2 A train applies its brakes to stop at a station and its speed (S km/h) is
   given by $S = 100 - 5t$ where t is the time in seconds after the brakes are
   applied.

t	5	0	4	8	12
S	5				

   a Copy and complete the table
      and draw a graph.

   b Find from your graph the speed
      when t equals
      i 2 seconds ii 10 seconds iii 16 seconds.

MyMaths.co.uk

Q 1059   SEARCH

Example

Marcus walks to work one morning
and the graph shows his progress.
He stops at a shop on the way.
Find from the graph

a   the distance from the shop to his
       workplace
b   the time he spends at the shop
c   his walking speed (in metres per
       minute) from his home to the shop.

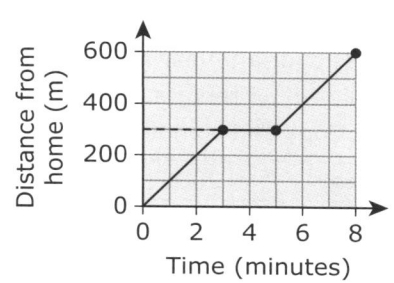

- -

a $600 - 300 = 300\,m$ **b** $5 - 3 = 2$ minutes
c $300 \div 3 = 100$ metres per minute

---

1   Mrs Mehta walks with her son to his school. She talks to a friend at the
       school gate and stops at her friend's house on the way home.

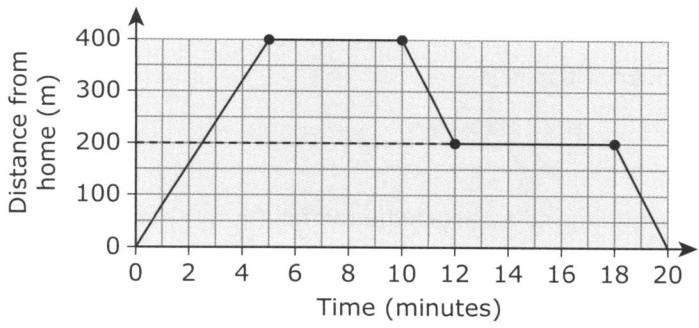

Find from the graph

a   the distance from the school to her friend's house
b   the time that she spends talking to her friend at the school gate
c   the time she spends at her friend's house
d   her walking speed (metres per minute) between
    i   her home and the school
    ii   her friend's house and her home.

The number of pupils who attend a nursery class over a 5-year period is shown on the graph.

a Copy and complete the table.

Year	2005	2006	2007	2008	2009
Number of pupils					

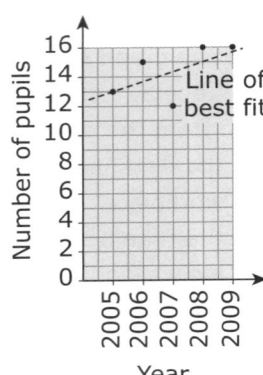

b Is the number showing a rising or falling trend?

c What is the mean number of pupils in the class over the 5-year period?

---

a

Year	2005	2006	2007	2008	2009
Number of pupils	13	15	12	16	16

b Rising, because the 'line of best fit' is sloping upwards.

c The mean number of pupils is

$$\frac{(13 + 15 + 12 + 16 + 16)}{5} = 72 \div 5 = 14.4$$

1 Joanne used a rain gauge to measure the rainfall (in mm) on each day of a certain week.

Day of the week	Mon	Tues	Wed	Thur	Fri	Sat	Sun
Rainfall (mm)	9	6	6	10	10	5	3

a Plot this information on a graph.

b Between which two days was the decrease the largest and what was this decrease?

c Did the rainfall show a rising or falling trend over the seven days?

d Find the mean daily rainfall for the week.

Example

Evaluate **a** 35.8 + 2.75   **b** 61 − 9.84

a Align the decimal points   **b** Align the decimal points

```
  35.8                                   61
  2.75 +                               9.84 −
  38.55                                51.16
```

1 Use the partitioning method to work out these.

a 8.8 + 3.4	**b** 9.5 + 4.6	**c** 7.5 + 5.7	**d** 6.8 + 2.5
e 9.9 + 6.6	**f** 7.2 + 1.8	**g** 9.8 − 3.1	**h** 8.1 − 4.3
i 7.6 − 5.8	**j** 9.3 − 6.4	**k** 6.5 − 1.7	**l** 7 − 1.1

2 Use the compensation method to work out these.

a 9.6 − 3.95	**b** 8.9 − 4.92	**c** 6.8 − 2.97	**d** 7.5 − 1.89
e 6.3 − 5.95	**f** 5.7 + 2.96	**g** 3.6 + 1.91	**h** 6.4 + 3.93

3 Work out these by any appropriate method.

Find which part, **i**, **ii** or **iii**, has a different answer from the other two.

a **i**   37.16 + 17.32 + 11.85   **b** **i**   40.36 + 19.29 + 12.91
    ii   21.43 + 32.81 + 12.1   **ii**   55.16 + 4.31 + 13.1
    iii   35.25 + 29.56 + 1.52   **iii**   45.32 + 12.25 + 15

c **i**   16.64 + 17.2 + 18   **d** **i**   35.97 − 16.23
    ii   16.5 + 3.35 + 32   **ii**   47.38 − 27.64
    iii   33.34 + 9.91 + 8.6   **iii**   51.27 − 31.52

4 a Find the perimeter of the triangle.

  b Find the fourth side of the trapezium
     if its perimeter is 27 cm.

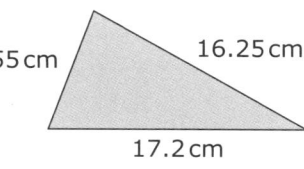

8.55 cm   16.25 cm

17.2 cm

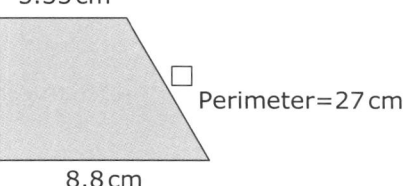

5.35 cm

6.15 cm

Perimeter = 27 cm

8.8 cm

Example

Work out 5.6 × 2.4

56 × 24 = 1344, but the product asked for will include two figures after the decimal point. Therefore the answer is 13.44. (There must be the same number of figures after the decimal point in the answer as in the starting numbers.)

1 Use the factors method to evaluate these.

a	32 × 0.03	**b**	21 × 0.04	**c**	11 × 0.05
d	26 × 0.02	**e**	33 × 0.03	**f**	34 × 0.02

2 Use the partitioning method to evaluate these.

a	1.2 × 32	**b**	2.3 × 13	**c**	3.3 × 21
d	3.2 × 33	**e**	2.1 × 43	**f**	1.3 × 31

3 Use any appropriate method to work these out.
Find which part **i**, **ii** or **iii**, has a different answer from the other two.

a	**i**	2.94 × 3	**ii**	1.75 × 5	**iii**	1.25 × 7
b	**i**	3.66 × 4	**ii**	2.44 × 6	**iii**	2.09 × 7
c	**i**	3.84 × 12	**ii**	3.36 × 14	**iii**	1.96 × 24
d	**i**	3.63 × 17	**ii**	2.37 × 26	**iii**	5.61 × 11
e	**i**	2.43 × 1.9	**ii**	1.71 × 2.7	**iii**	1.65 × 2.8
f	**i**	3.42 × 2.1	**ii**	2.98 × 2.4	**iii**	5.13 × 1.4

4 Find the area of each of these shapes.

a

6.25 m

4.8 m

b

0.45 m

0.32 m

c

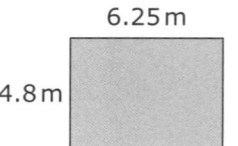

30.5 cm

←16 cm→

Example

Evaluate

a $420 \div 8$ **b** $36 \div 1.5$

- -

a $420 \div 8$

$= (400 \div 8) + (20 \div 8)$

$= 50 + 2\,R4$

$= 52 + (4 \div 8)$

$= 52.5$

b $36 \div 1.5 = \dfrac{36}{1.5} = \dfrac{360}{15} = \dfrac{72}{3} = 24$

1 Use the factors method to evaluate these.

a	$378 \div 18$	**b**	$864 \div 12$	**c**	$972 \div 27$	**d**	$434 \div 14$
e	$903 \div 21$	**f**	$975 \div 15$	**g**	$902 \div 22$	**h**	$768 \div 16$

2 Use the partitioning method to evaluate these.

a	$648 \div 12$	**b**	$780 \div 15$	**c**	$336 \div 16$	**d**	$325 \div 25$
e	$468 \div 15$	**f**	$279 \div 12$	**g**	$516 \div 16$	**h**	$230 \div 25$

3 An electric train is 126 m long and it consists entirely of vehicles which are 10.5 m long.

How many vehicles does it consist of?

4 Hazel Avenue is 100 m long and lamp posts are arranged as shown with equal spaces of 12.5 m between them. How many lamp posts are there on one side of the road?

Hazel Avenue

$\longrightarrow$ 12.5 m $\longleftarrow$

$\longleftarrow$ 100 m $\longrightarrow$

Work out   **a** $52 + 15 \times 12$ **b** $\sqrt{(61^2 - 11^2)}$

a $52 + 15 \times 12 = 52 + 180 = 232$
    (Note that the multiplication has to be done first,
    because there are no brackets – remember BIDMAS.)

b $\sqrt{(61^2 - 11^2)} = \sqrt{(3721 - 121)} = \sqrt{3600} = 60$

1   Work these out. Find which part **i**, **ii** or **iii**, has a different answer
    from the other two.

	i	**ii**	**iii**
a	$56 + 39 \times 12$	$66 + 33 \times 14$	$92 + 27 \times 16$
b	$512 - 28 \times 15$	$496 - 25 \times 16$	$558 - 22 \times 21$
c	$192 - 72 \div 12$	$205 - 812 \div 28$	$201 - 550 \div 22$
d	$75 + 408 \div 17$	$67 + 899 \div 29$	$72 + 648 \div 24$
e	$(520 - 304) \div 9$	$(305 - 109) \div 7$	$(271 - 79) \div 8$
f	$(100 - 81) \times 12$	$(150 - 129) \times 11$	$(96 - 85) \times 21$

Work out each part for questions **2**, **3** and **4**.

2   **a** $\dfrac{(9^2 + 7^2)}{\sqrt{25}}$ **b** $\dfrac{(10^2 + 8^2)}{\sqrt{16}}$ **c** $\dfrac{(12^2 + 8^2)}{\sqrt{64}}$ **d** $\dfrac{(17^2 - 3^2)}{\sqrt{49}}$

3   **a** $\sqrt{(41^2 - 40^2)}$ **b** $\sqrt{(29^2 - 20^2)}$ **c** $\sqrt{(53^2 - 28^2)}$ **d** $\sqrt{(15^2 + 8^2)}$

4   **a** $\dfrac{(4^2 + 3^2)(5^2 + 2^2)}{(3^2 - 2^2)}$ **b** $\dfrac{(6^2 + 5^2)(7^2 - 3^2)}{(26^2 - 24^2)}$

     c $\dfrac{(10^2 - 7^2)(9^2 - 3^2)}{(7^2 - 3^2)}$ **d** $\dfrac{(9^2 + 5^2)(11^2 - 3^2)}{(7^2 + 2^2)}$

5   Find the area of the yard.

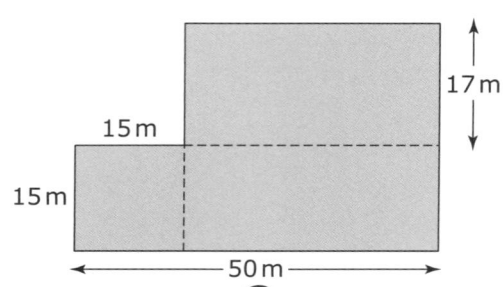

MyMaths.co.uk

Q 1167   SEARCH

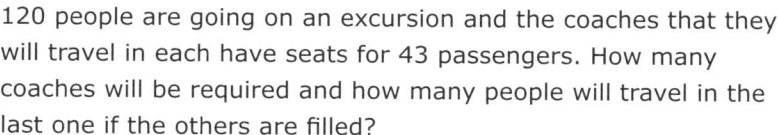

Example

120 people are going on an excursion and the coaches that they will travel in each have seats for 43 passengers. How many coaches will be required and how many people will travel in the last one if the others are filled?

$\frac{120}{43}$ = 2 remainder 34. Therefore 3 coaches are required and 34 people will travel in the third one if the others are filled.

For questions **1** and **2** express your answer as a fraction in its lowest terms.

1 Candace's car travels 18 km on one litre of petrol. Find how many litres it will require for a journey of

  a 78 km   **b** 120 km   **c** 165 km   **d** 51 km.

2 Jisanne's bike has wheels with a circumference of 210 cm. How many revolutions will they make when she rides 175 m from her house to the post office?

3 A ferry operator's boat can accommodate 14 passengers. One day he unexpectedly finds 75 people waiting. How many crossings will he have to make and how many people will be in his boat for the last crossing?

4 A train has 1200 tonnes of coal to move and each of its wagons can hold 150 tonnes. How many wagons will be required and how many tonnes will the last one carry if all of the others are full?

5 At Holly Hill School the lunchtime break lasts for 95 minutes. Six teachers share the supervision for equal times and the time left over is covered by the Deputy Head. How long will each teacher spend on duty and for how many minutes will the Deputy Head have to supervise at the end?

Example

Yasirah thinks that many older people will be quite happy if the government raises the retiring age, because they will need more money if they are expected to live longer. She asked the older workers at a factory gate what they think on their way into work, and most of them seemed to agree with her. Is she collecting primary or secondary data? Why, however, could her results be misleading?

She is collecting primary data because she is asking the workers themselves. Her results may be misleading because she has only asked factory workers. Other kinds of workers may think differently.

1   Keanu thinks that in European football matches Italian clubs play defensively. How could he collect any data to test this claim? Would the data be primary or secondary?

2   Naim lives near to the Yorkshire coast and it is well known that the sea is encroaching on the land there.
Naim, however, thinks that the rate of encroachment is slowing down. He finds details that tell him the distance between a very solid rock and the high tide water position on the same date of each year over a period of many years. Are these primary or secondary data? The details suggest that he is right, but why might this conclusion be misleading? (Look carefully at the illustration.)

Distance recorded

3   Jodie runs a cat boarding house and she thinks that cats eat less as they get older. She decides to collect some primary data by recording the daily food consumption of several cats of different ages. Suggest two important things that she must do.

Example

Zoey is doing a survey for a train operator to see how passengers who use a certain commuter train think that their service might be made better. One morning she walked through the train and asked 100 passengers what improvements they would like to see, but the answers she received were vague. Suggest how she might obtain useful answers.

- -

She could ask every passenger to fill in a questionnaire sheet such as this one.

Service that needs improving?

Tick the appropriate column(s).	Punctuality	Better tannoy information	Comfort of seats	Manners of railway staff	Heating	Lighting

1 Xiao-Li is trying to find out whether or not pupils at her school spend more time on computers as they get older. She makes a question sheet like the one below for each of the years 7, 8, 9, 10 and 11, then asks 20 pupils from each year to write in their names and tick the appropriate column.

Name	Time spent on a computer per day					
	None	0 to 1 h	1 to 2 h	2 to 3 h	3 to 4 h	More than 4 h
Tim Jones						

Before she begins to draw conclusions, what is one very important thing that she should do with the results?

2 The Principal of a tertiary college is interviewing students from local schools who want to continue their studies at his college. He decides that he would like a simple questionnaire sheet for them all to fill in which will tell him all the most important things about the student. Design a sheet which you think would be appropriate for him to use.

Example

One week these were Suzanne's times for travelling to work on the bus.
34, 32, 34, 31, 32 and 33 minutes
Copy and complete this frequency table.

Time (min)	Tally	Frequency
31		
32		
33		
34		

Time (min)	Tally	Frequency
31	I	1
32	II	2
33	I	1
34	II	2

1 These are the numbers of children who travelled on the school bus to Ash Lane School over a 30-day period.

27 26 28 27 29 25 28 25 29 26 30 30 28 28 27
28 29 27 28 26 27 28 24 29 27 30 29 26 30 25

Display the details on a tally and frequency table.

2 These are the best times for the girls in Class 9B for the 100 m event.

13.1, 13.3, 14.9, 12.7, 13.4, 14.2, 14.1, 12.3, 13.0, 14.3, 14.1, 12.6, 14.0, 13.2, 14.4, 13.1, 13.2, 14.6, 13.3 and 12.8 seconds

Display the details on a tally and frequency table.
(Use class intervals of 12.0 to 12.4, 12.5 to 12.9, 13.0 to 13.4, 13.5 to 13.9, 14.0 to 14.4 and 14.5 to 14.9 s.)

Example

a The pie chart shows the proportion of pupils who walk, cycle or travel by bus to High Park School. If there are 300 pupils altogether and none travel by any other means, find the numbers for the three ways of travelling.

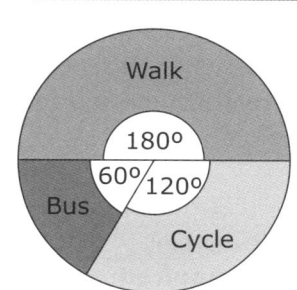

b Display your details on a pictogram.

a The number who walk is $\frac{180}{360} \times 300 = 150$

The number who cycle is $\frac{120}{360} \times 300 = 100$ and the number who bus is $\frac{60}{360} \times 300 = 50$

b

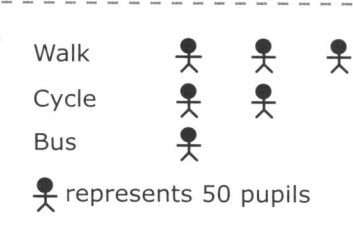

1 The pictogram shows how many people used a ferry on the four days of an Easter weekend.
Copy and complete the table and display the details on a pie chart.

Friday 👤 👤
Saturday 👤 👤 👤
Sunday 👤 👤 👤 👤
Monday 👤 👤 👤

👤 represents 60 people

Day	Number of people	Total number for the weekend	Angle required
Friday			
Saturday			
Sunday			
Monday			

2 Where Rapinda lives the temperature can change very quickly. The frequency table summarises the midday temperature over a 2-week period. Display the details on a frequency diagram.

Temperature (°C)	Frequency
0 up to 5	2
5 up to 10	5
10 up to 15	6
15 up to 20	1

Example

Jim is a forester and he thinks that conifer trees grow slower at higher altitudes. He checks this by planting four similar trees at places of different altitude and seeing how long they each take to grow 1 m tall.

Draw this data on a scatter graph. Do you think that Jim is correct

Altitude (m)	100	200	300	400
Time required (months)	20	17	19	16

Time required (months) / Altitude (m) — Line of best fit

Yes, because the 'line of best fit' is a falling line.

---

1 In a 5000 m race the runners have a 200 m 'run in' followed by 12 complete laps. Anwar thinks that his lap times decrease towards the end of the race. His friend recorded his lap times for a certain race.

Number of laps	1	2	3	4	5	6	7	8	9	10	11	12
Lap times (s)	68	70	72	68	66	64	68	66	66	68	66	60

Draw these data on a scatter graph. Do you think from this that Anwar is correct?

2 In high jump contests the competitors are allowed three attempts to clear the bar at each height. Surprisingly, Peter thinks that he clears the bar after fewer attempts as the height of the bar is raised. Here are his details for a contest which he won.

Height of the bar (cm)	140	145	150	155	160	165	170	175
Number of attempts	2	3	1	2	3	1	1	2

Draw this data on a scatter graph. Do these data suggest that Peter is right?

Example

Layla recorded the temperature outside her house at midday on each day of a certain week. The figures were 12, 11, 10, 11, 8, 7 and 11°C. Find

a the mode **b** the median **c** the mean for these temperatures.

- -

a Arranging the temperatures in ascending order gives
7, 8, 10, <u>11</u>, 11, 11 and 12°C.

11°C is the mode because it is the temperature that occurs most frequently.

b 11°C is also the median because it is the middle temperature in the list. (It is underlined to indicate its position.)

c The mean = total of the terms ÷ number of terms
= (12 + 11 + 10 + 11 + 8 + 7 + 11) ÷ 7 = 70 ÷ 7 = 10°C

For all questions find **a** the mode **b** the median **c** the mean.

1 Klara cycles to school and these were her times in minutes over a 2-week period. (Friday at the end of the second week was a school holiday.)

Mon	Tues	Wed	Thur	Fri	Mon	Tues	Wed	Thur
17	19	20	19	21	20	19	22	23

2 Here is a train's punctuality record for a certain week.

	Mon	Tues	Wed	Thur	Fri	Sat
Minutes late	2	0	3	0	54	1

Why is the mean a misleading average for these data?

3 Over a 2-week period these were the numbers of students who used the school bus to travel to West Hill School.

Mon	Tues	Wed	Thur	Fri	Mon	Tues	Wed	Thur	Fri
23	15	19	22	21	15	19	15	24	23

Why is the mode a misleading average for these data?

1 The students from four schools, North Mead, Oak Tree Lane, Larch Avenue and Middle Street, all go to the town hall to hear four singers, Julian, Wayne, Zoey and Charmaine, perform. They all voted for the singer that they thought was the best and the bar charts show their results.

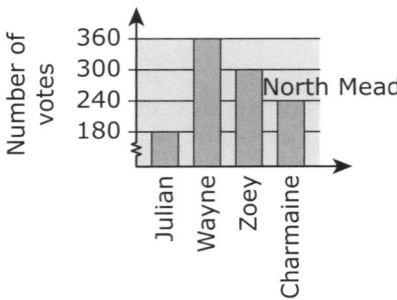

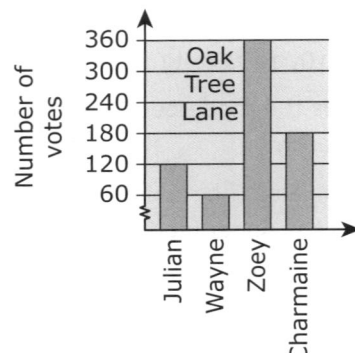

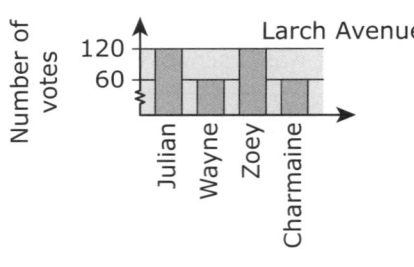

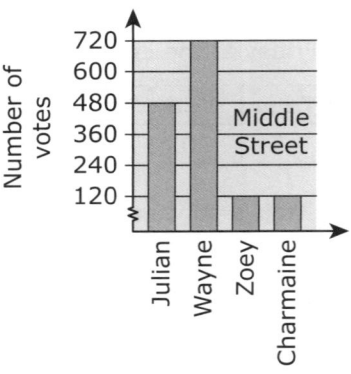

a Work out which singer got the most votes.

b Display the total number of votes on a graph or chart of your choice. (You could use a pie chart, bar chart ...)

Example

An office manager thinks that the staff in his office who have longer travelling times to work suffer more stress and that this leads to them having more days off work. Here are the details for the office staff in 2008.

	Bob	Amir	Rose	Jasmine	Shimi
Travelling time to work (min)	20	70	40	90	10
Days off in 2008	5	8	4	4	1

Plot these details on a scatter graph and comment on the correlation.

- -

The correlation is positive because the 'line of best fit' is a rising line. The manager, therefore, appears to be right.

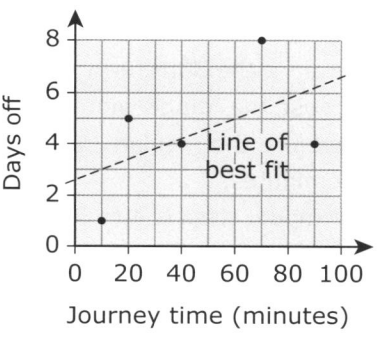

1 Five people are moving because their company has had to redeploy staff. Here are the details of their annual salaries and the selling values for their houses.

	Georgia	Greg	Rani	Candace	Shani
Annual salary (£)	25000	17000	32000	21000	12000
House valuation (£)	500000	440000	480000	360000	360000

Plot these figures on a scatter diagram and comment on the correlation.
Does it suggest that higher earners live in more expensive houses?

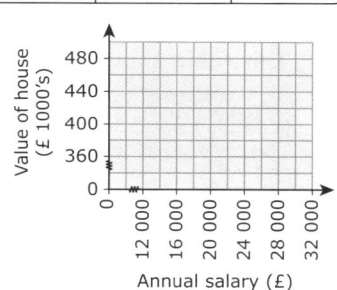

Example

Martin recorded the lateness of his school bus over 60 days.

Lateness	0–3 min	4–7 min	8–11 min
Number of times	20	32	8

Find   a   the mean   b   the median
        c   the modal class   d   the range for these data.

a   The mean = (1.5 × 20 + 5.5 × 32 + 9.5 × 8) ÷ 60
              = 282 ÷ 60 = 4.7 minutes

b   The median = 30th term. There are 20 terms up to 4 minutes,
     therefore the 30th term equals 4 + (10 terms out of 32) × 3,
     because the 'range widths' are each 3 minutes.
     Therefore the median = $4 + \frac{10}{32} \times 3 = 4 + \frac{30}{32} = 4.9375$ or approximately
     4.9 minutes.

c   The modal class is 4 to 7 minutes, the class with the highest
     frequency.

d   The range for the whole data is 11 − 0 = 11 minutes.

All questions refer to the 120 students in Year 9 at Hazel Avenue School.
For all questions find
a   the mean   b   the median
c   the modal class   d   the range for the data.

1   The table shows their travelling times on their way to school.

Time (min)	0–4	5–9	10–14	15–19	20–24
Number of students	20	36	40	14	10

2   This table shows how many hours they watch television per week.

Time (hours)	0–4	5–9	10–14	15–19	20–24	25–29
Number of students	15	18	20	32	20	15

3   The table shows their weekly earnings from part-time work.

Amount earned (£)	0–19	20–39	40–59	60–79
Number of students	20	20	50	30

MyMaths.co.uk

Q 1201   SEARCH

Example

Eva and Simone played in four matches for a cricket team.
Their batting scores were these. Find the mean score for both and
comment on their performances.

Eva	13	7	21	0	5	15	9	2
Simone	21	50	0	7	9	0	17	0

For Eva's results the mean =
(13 + 7 + 21 + 0 + 5 + 15 + 9 + 2) ÷ 8 = 72 ÷ 8 = 9
For Simone's results the mean =
(21 + 50 + 0 + 7 + 9 + 0 + 17 + 0) ÷ 8 = 104 ÷ 8 = 13

Therefore Simone's average record is the better but she is more
likely to be 'out for a duck'.

Her range is much larger than Eva's, 50 − 0 = 50 as opposed to
21 − 0 = 21.

1 Josiah and Tony both play for the same football team and their goal
scoring records for the 2008 season are shown on the bar charts.

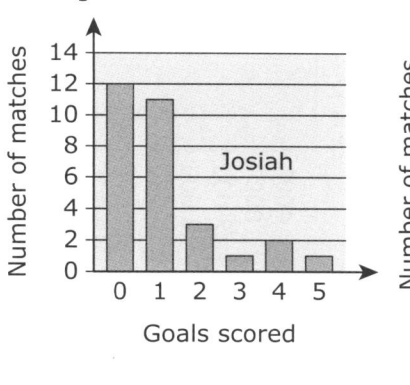
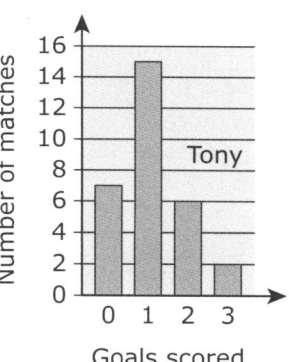

a Find **i** the mean **ii** the mode **iii** the range for each player.
b What do the modes and the ranges tell you about the difference(s)
in their records? Explain your answer(s).

⊛ MyMaths.co.uk

Example

The table is a simplified version of Jenny's geography results which are shown in your students' book.

Country	Mean annual income (US $)	Population growth (%)	Adult mortality	Life expectancy (years)	Number of doctors per 1000 people
Angola	4000	2.8	500	40	0.1
South Africa	9000	0.7	560	50	0.7
Brazil	9000	1.3	180	70	1.0
Chile	11000	1.0	90	80	1.0
Russia	13000	−0.5	300	70	4.3
Germany	33000	0.0	80	80	3.4

Draw a scatter diagram of doctors per 1000 people against mean annual income. Comment on the correlation and give any reason you can think of for why it might be so.

The correlation is positive because the 'line of best fit' is a rising line. The correlation is expected because a higher income means that more doctors can be paid.

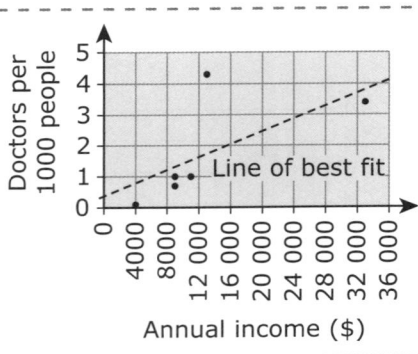

For both questions use the table given in the example above to draw the scatter diagram asked for. Comment on the correlation for each case and give any reason(s) for the correlation that you can think of.

1   Draw a scatter diagram of life expectancy against doctors per 1000 people.

2   Draw a scatter diagram of population growth against doctors per 1000 people.

Example

a Draw a reflection in the *x*-axis
of the triangle O.

b Draw an anticlockwise rotation
of the triangle O about (0, 0).

The images are shown
on the diagram.

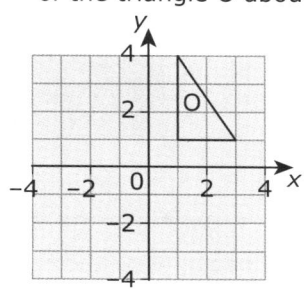

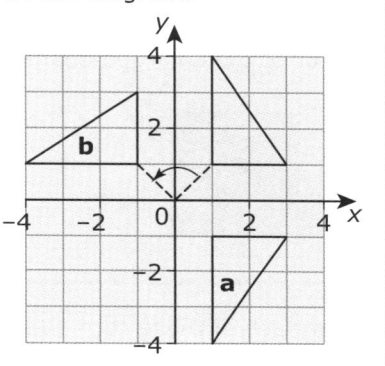

1 a Draw a copy of the diagram
illustrated. Draw these images
on your copy.

i Reflect O in the *x*-axis.
Label the image A.

ii Reflect O in the *y*-axis.
Label the image B.

iii Reflect A in the *y*-axis.
Label the image C.

iv Reflect O in the line
$y = x$. Label the image D.

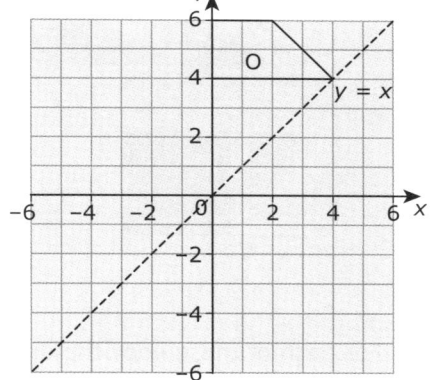

v Reflect D in the *x*-axis. Label the image E.

vi Reflect E in the *y*-axis. Label the image F.

vii Reflect F in the *x*-axis. Label the image G.

b What rotation maps

i O onto C   **ii** O onto G   **iii** O onto E

iv D onto B   **v** D onto F   **vi** D onto A?

c Look at the pattern you have drawn. State

i the number of lines of symmetry it has and where they are

ii its order of rotational symmetry.

Example

Copy the diagram onto squared paper and mark the centre of enlargement. State the scale factor of the enlarged image.

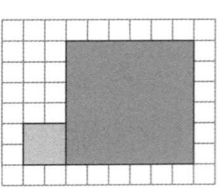

The dot marks the position of the centre of enlargement. The scale factor is 3.

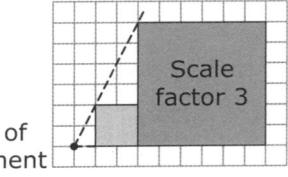

Scale factor 3

Centre of enlargement

1 For each of these copy the diagram onto squared paper. Mark the centre of enlargement and state the scale factor for each enlarged image.

a **b**

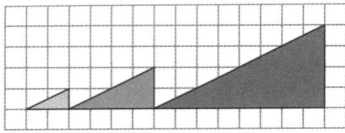

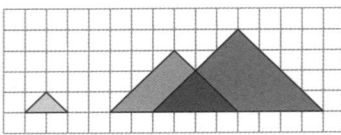

c **d**

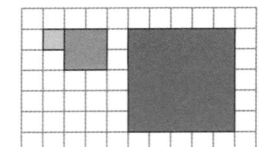

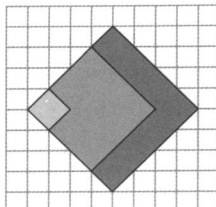

2 For each of these copy the diagram onto squared paper and draw on the images whose scale factors are given.

a **b** **c** 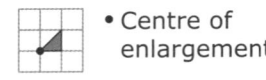 • Centre of enlargement

a Scale factor of **i** 2 **ii** 4 **iii** 6
b Scale factor of **i** 2 **ii** 3 **iii** 5
c Scale factor of **i** 2 **ii** 4 **iii** 5

3 Work out the missing side lengths.

7.5 cm

10 cm

3 cm Scale factor $1\frac{1}{2}$

Scale factor $2\frac{1}{2}$

Example

Look at the diagram. The two triangles are similar in shape. Find the centre of enlargement and state the scale factor of the smaller triangle with respect to the larger one.

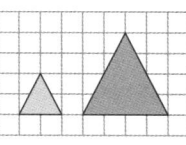

The scale factor is $\frac{1}{2}$ because its side lengths are half of the corresponding ones on the larger triangle. The centre of enlargement is marked with a dot.

Centre of enlargement

Scale factor $\frac{1}{2}$

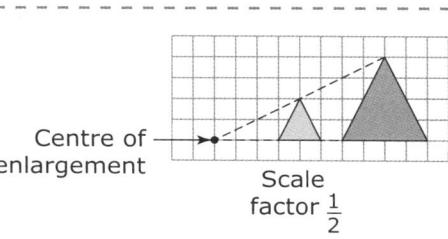

1 For these copy the diagram, mark on the centre of enlargement and state the scale factor of each figure with respect to the largest one.

a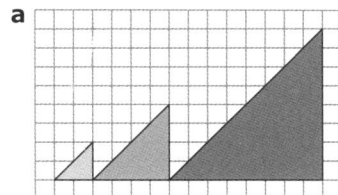

b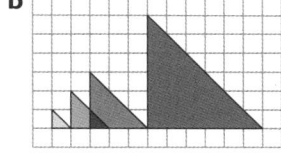

c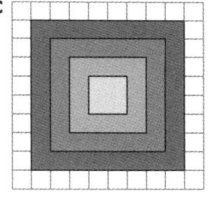

2 For these copy the diagram and draw on each figure the images whose scale factors are given. Each centre of enlargement is marked with a dot.

a Scale factor of   i $\frac{1}{2}$ ii $\frac{1}{4}$ b Scale factor of   i $\frac{1}{2}$ ii $\frac{1}{4}$ iii $\frac{3}{4}$

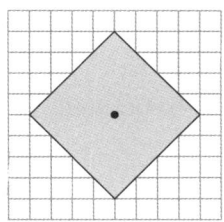

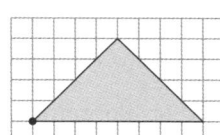

c Scale factor of   i $\frac{1}{2}$ ii $\frac{1}{3}$ iii $\frac{2}{3}$

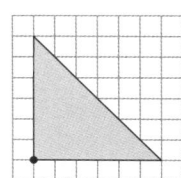

Example

On a village plan the distance between the post office and the church is 12.5 cm. If the scale of the plan is 1 cm represents 20 m, what is the real distance between the two?

The real distance = 12.5 × 20 m = 250 m

1 Jordan has a model railway set which includes a scale replica of his local station. His models are all made with a scale of 1 : 75 with respect to their real counterparts. Copy and complete the table.

	Dimension on the real railway (m)	Dimension on Jordan's model (cm)
Length of the station platform	54	
Height of the footbridge	6	
Width of the station platform	9	
Height of the station lamp posts	2.4	
Length of a coach		22
Width of the same coach		5.2
Height of the same coach		3.9
Track gauge		1.92

2 Tim looks at a plan of the town where he lives and notices that the distance on it between his house and his school is 50 cm. The distanceometer on his bike, however, measures the real distance to be 2 km. What is the scale of the plan?

3 Tnisha walks in a straight line for 105 m. She turns through 90° and walks for a further 60 m.
   a Draw a scale drawing of her walk using a scale of 1 cm represents 10 m.
   b Find her direct distance from her starting point.

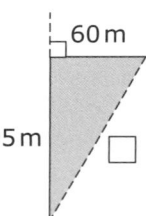

60 m

105 m

MyMaths.co.uk

Q 1103, 1117 **SEARCH**

Example

Ambleside is 7 km from
Grasmere on a bearing of
135°. Draw this on a scale
diagram using a scale of
1 cm represents 2 km.

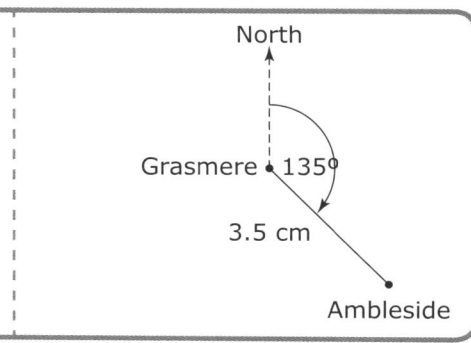

1 Study this diagram
   carefully about the
   positions of places
   in a town. Copy
   and complete the
   table.

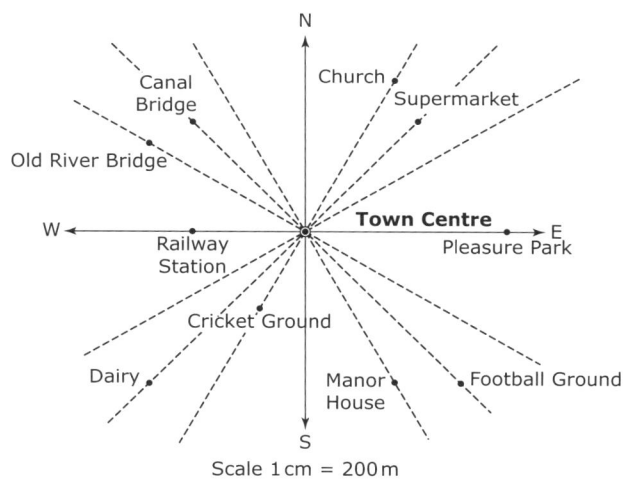

Scale 1 cm = 200 m

Place	Bearing from centre	Distance from centre	Place	Bearing from centre	Distance from centre
Church			Cricket Ground		
Supermarket			Dairy		
Pleasure Park			Railway Station		
Football Ground			Old River Bridge		
Manor House			Canal Bridge		

Solve these equations: **a** $2x + 3 = 11$ **b** $\dfrac{(y-1)}{5} = 3$

---

a $2x + 3 = 11$, therefore $2x = 11 - 3 = 8$, so $x = 8 \div 2 = 4$

b $\dfrac{(y-1)}{5} = 3$, therefore $y - 1 = 3 \times 5 = 15$, so $y = 15 + 1 = 16$

For questions **1** to **4** solve the equations by any appropriate method. Find which equation, **a**, **b** or **c**, has a different solution from the other two.

1 **a** $x + 5 = 8$ **b** $x + 11 = 15$ **c** $x + 9 = 12$

2 **a** $y - 5 = 12$ **b** $y - 9 = 7$ **c** $y - 12 = 4$

3 **a** $5z = 65$ **b** $4z = 60$ **c** $7z = 105$

4 **a** $\dfrac{t}{3} = 16$ **b** $\dfrac{t}{8} = 6$ **c** $\dfrac{t}{7} = 7$

5 Solve these equations which have negative solutions.

a $2x + 11 = 5$ **b** $6y + 19 = 1$ **c** $\dfrac{z}{3} + 5 = 2$

d $\dfrac{t}{4} + 7 = 2$ **e** $\dfrac{(u+10)}{2} = 4$ **f** $\dfrac{(v+15)}{3} = 3$

6 **a** This isosceles trapezium has a perimeter of 26 cm.
Find the value of x.

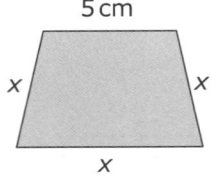

b If the distance from London to Exeter is 273 km, find the value of x and the distance from Basingstoke to Exeter.

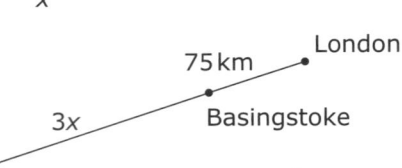

c This isosceles triangle has a perimeter of 36 cm.
Find the three side lengths.

MyMaths.co.uk

Q 1154 SEARCH

Solve for x the equation $3(2x + 3) - 2(2x - 1) = 17$

$3(2x + 3) - 2(2x - 1) = 17$
Therefore $6x + 9 - 4x + 2 = 17$
therefore $\qquad 2x + 11 = 17$
therefore $\qquad\qquad 2x = 6$
therefore $\qquad\qquad\quad x = 3$

1 Solve these equations.

  a $2(3x + 4) = 26$ **b** $2(5y + 2) = 34$ **c** $4(2v - 7) = 36$

  d $3(3m - 2) = 30$ **e** $\frac{1}{2}(4n - 6) = 7$ **f** $\frac{1}{3}(2p - 7) = 9$

2 Solve these equations which have negative solutions.

  a $2(3x + 17) = 4$ **b** $3(4y + 19) = 33$

  c $4(2z + 13) = 28$ **d** $5(2t + 15) = 35$

3 Solve these equations.

  a $3(2x + 5) + 2(5x + 3) = 53$ **b** $4(3y + 2) + 5(2y + 3) = 111$

  c $4(3z + 4) + 3(2z - 3) = 61$ **d** $2(7t + 8) + 3(2t - 1) = 93$

  e $3(5u - 2) + 4(2u + 7) = 68$ **f** $5(3v - 7) + 2(3v + 2) = 74$

4 Solve these equations. Take care with multiplying positive and negative numbers.

  a $5(3x + 7) - 3(2x + 3) = 53$ **b** $7(2y + 3) - 2(4y + 7) = 67$

  c $2(6z + 5) - 3(2z - 3) = 61$ **d** $3(4t + 1) - 2(5t - 2) = 25$

  e $5(2u - 3) - 2(4u + 5) = 9$ **f** $5(3v - 2) - 3(4v + 3) = 44$

5 The signboard illustrated has a perimeter of 3.4 m.

Find the value of x and the five dimensions of the signboard.

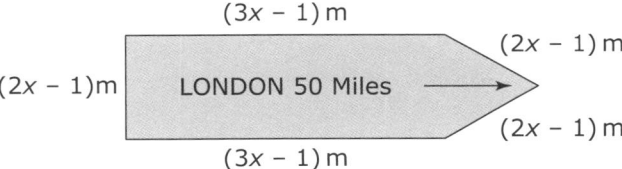

$(3x - 1)$ m

$(2x - 1)$ m

$(2x - 1)$m   LONDON 50 Miles

$(2x - 1)$ m

$(3x - 1)$ m

Example

Solve for x the equation $\dfrac{(4x - 7)}{5} = \dfrac{(7x - 11)}{9}$

$\dfrac{(4x - 7)}{5} = \dfrac{(7x - 11)}{9}$

Therefore $9(4x - 7) = 5(7x - 11)$

therefore $\quad 36x - 63 = 35x - 55$

therefore $36x - 35x = 63 - 55$

therefore $\qquad\qquad x = 8$

For questions **1**, **2** and **3** solve the equations.

1 **a** $5x + 2 = 2x + 11$ **b** $7y + 3 = 3y + 19$

   c $10z + 1 = 4z + 13$ **d** $8t - 7 = 3t + 8$

   e $9u - 11 = 6u + 13$ **f** $10v - 13 = 7v + 5$

2 **a** $15x + 7 = 2(3x + 26)$ **b** $11y + 3 = 4(2y + 3)$

   c $15z - 2 = 4(2z + 3)$ **d** $25p - 9 = 6(3p + 2)$

   e $11q - 29 = 3(2q - 3)$ **f** $15m - 41 = 4(2m - 5)$

3 **a** $\dfrac{(13x + 5)}{3} = 3x + 7$ **b** $\dfrac{(10y + 3)}{3} = 2y + 9$

   c $\dfrac{(13m - 11)}{5} = 2m + 5$ **d** $\dfrac{(15p - 8)}{2} = 6p + 5$

   e $\dfrac{(13q - 35)}{3} = 3q - 5$ **f** $\dfrac{(19x - 29)}{5} = 3x - 5$

   g $\dfrac{(6y + 7)}{5} = \dfrac{(3y + 11)}{4}$ **h** $\dfrac{(3z + 13)}{5} = \dfrac{(5z + 3)}{6}$

4   The isosceles trapezium and the isosceles triangle have the same
    perimeter. Find the value of x and the side lengths for both figures.

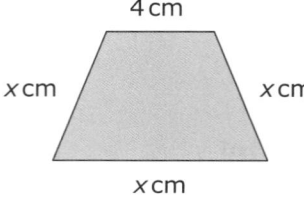

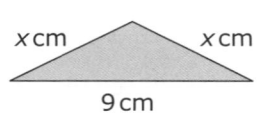

MyMaths.co.uk

Q 1182   SEARCH

10d Constructing equations

Louis has n marbles. He wins a game and finds that he has twice as many, but loses 5 in another game. If he now has 13 marbles, find his original number (n).

- -

The equation to solve is $2n - 5 = 13$ (n is multiplied by 2 and then 5 is subtracted.)

If $2n - 5 = 13$

then $2n = 13 + 5$

so $2n = 18$

hence $n = 9$

1 a If the perimeter of the isosceles triangle is 40 cm, find the value of d and the three side lengths of the triangle.

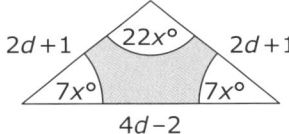

$2d+1$ $22x°$ $2d+1$

$7x°$ $7x°$

$4d-2$

b Find the value of x and the sizes of the angles in the triangle.

2 Scafell Pike, England's highest mountain, is 979 m high. Carrauntoohil, Ireland's highest mountain, is x m higher and Ben Nevis, Scotland's highest mountain, is $6x$ m higher. If the height of Ben Nevis is 1345 m find **a** the value of x **b** the height of Carrauntoohil.

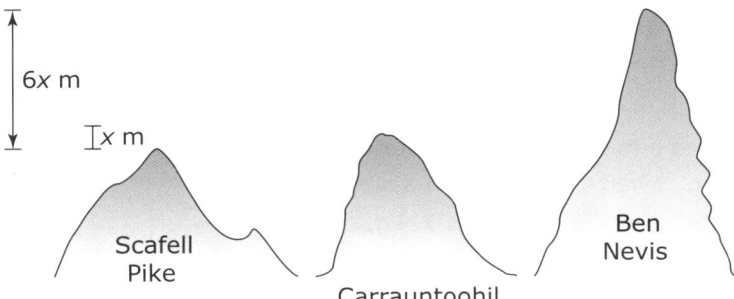

$6x$ m

x m

Scafell
Pike

Carrauntoohil

Ben
Nevis

Example

Solve the equation $x^2 + x = 11$ by trial and error, giving your answer correct to one decimal place. (Use $x = 3$ for the first trial.)

$x^2 + x = 11$, but $3^2 + 3 = 9 + 3 = 12$, which is too high,
and $2.9^2 + 2.9 = 8.41 + 2.9 = 11.31$, which is still too high,
but $2.8^2 + 2.8 = 7.84 + 2.8 = 10.64$, which is too low.
11.31, however, is nearer to 11 than 10.64, so 2.9 is the solution correct to one decimal place.

1 Using $x = 9$ for the first trial, find a solution for these correct to one decimal place.

   a $x^2 + x = 90$ **b** $x^2 + x = 85$ **c** $x^2 + x = 75$

   d $x^2 + x = 80$ **e** $x^2 + x = 100$ **f** $x^2 + x = 105$

2 Using $x = 12$ for the first trial, find a solution for these correct to one decimal place.

   a $x^3 + x = 1728$ **b** $x^3 + x = 1650$ **c** $x^3 + x = 1600$

   d $x^3 + x = 1700$ **e** $x^3 + x = 1800$

3 Solve these by trial and error. Choose your own number for your first trial.

   a $x^2 + 2x = 360$ **b** $x^2 + 5x = 750$ **c** $y^2 - 3y = 270$

   d $y^2 - 4y = 480$ **e** $(x + 1)(x + 2) = 72$ **f** $(x + 1)(x + 3) = 63$

4 A yard has a length which is 3 m longer than its width.
If its area is 180 m², find its two dimensions.

$(x+3)$m

x m   Area = 180 m²

5 Eva's baby has a mass of m kg and her own mass is m^2 kg. When she stands on a scale and is holding her baby the reading is 72 kg.
Find the mass for each of them on their own.

Find **a** the square root **b** the cube root of 64.

a $\sqrt{64}$ = 8 because 8 × 8 = 64

b $\sqrt[3]{64}$ = 4 because 4 × 4 × 4 = 64

1 Find the square roots of these numbers using factors.

a 900	**b** 3600	**c** 6400	**d** 4900	**e** $\frac{16}{100}$
f $\frac{25}{100}$	**g** $\frac{64}{100}$	**h** 0.0001	**i** 256	**j** 324

2 Use any convenient method to find the square roots of these numbers correct to one decimal place.

a 20	**b** 12	**c** 35	**d** 44
e 110	**f** 240	**g** 200	**h** 180

3 Find the cube roots of these numbers by any convenient method.

a 8000	**b** 125 000	**c** 27 000	**d** 512
e 216	**f** 729	**g** 343	**h** 1728

4 Use any convenient method to find the cube roots of these numbers correct to one decimal place.

a 20	**b** 25	**c** 50
d 100	**e** 35	**f** 10

5 This square yard has an area of 729 m². Find its side length.

Area = 729m²

6 If A is the area of a circle, the radius r is given (very nearly) by $r^2 = \left(\frac{7A}{22}\right)$. Use this formula to find the radius of the circular pond of area 38.5 m².

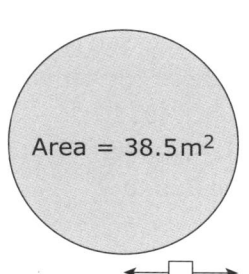

Area = 38.5m²

Example

Simplify $\dfrac{10(x^2)^3}{5x^4}$

$\dfrac{10(x^2)^3}{5x^4} = 10x^6 \div 5x^4 = 2x^2$

1 Simplify these expressions using indices in your answers.

   a $x \times x \times x \times x$

   b $y \times y \times y \times y \times y \times y$

   c $x \times x \times x \times x \times x \times x \times z \times z$

   d $3 \times 3 \times 3 \times 3 \times q \times q \times q$

2 Simplify these expressions.

   a $x^3 \times x^4$ **b** $y^5 \times y^6$ **c** $z^9 \times z^7$

   d $u^4 \times u$ **e** $v^3 \times v^5 \times v$ **f** $p^3 \times p^4 \times q^2 \times q^6$

3 Simplify these expressions.

   a $x^9 \div x^4$ **b** $y^{10} \div y^6$ **c** $z^6 \div z^3$ **d** $m^7n^5 \div m^2n^3$

   e $p^5q^6 \div p^2q^3$ **f** $m^6n^3 \div m^2n$ **g** $s^6t^5 \div s^3t^4$ **h** $y^6z^3 \div y^2z^3$

4 Simplify these bracketed expressions.

   a $(x^3)^2$ **b** $(y^5)^3$ **c** $(z^2)^5$ **d** $(t^4)^4$

   e $(x^2y^3)^4$ **f** $(m^3n^2)^3$ **g** $(p^2q^4)^2$ **h** $(r^3s)^5$

5 Simplify these expressions. Simplify both the numbers and the letters where required.

   a $3x^2 \times 4x^7$ **b** $5y^2 \times 3y^4$ **c** $\dfrac{12z^7}{3z^3}$ **d** $\dfrac{t^3 \times t^9}{t^5}$

   e $\dfrac{u^2 \times u^4}{u^3}$ **f** $\dfrac{v^3 \times v^8}{(v^2 \times v^5)}$ **g** $\dfrac{w^2 \times w^8}{(w^5 \times w^3)}$ **h** $\dfrac{15a^3 \times a}{3a^5}$

   i $\dfrac{20b^5 \times b^2}{4b^4}$ **j** $\dfrac{30a^4 \times c}{5c^2}$

6 If the distance from the Sun to the planet Pluto is written as $6x^9\,$km, then the distance from the Sun to the Earth is $1.5x^8\,$km.

   a Find the ratio of these two terms.

   b If the ratio is in fact $40:1$, find the value of x and the distance from the Sun to each planet.

☀ Sun   ● Earth   Pluto ●

← $1.5x^8$ km →

$6x^9$ km

MyMaths.co.uk

Q 1033   SEARCH

Example

a Simplify $\sqrt{500}$ as much as possible.
b Write $\sqrt[3]{15}$ in index form.

a $\sqrt{500} = \sqrt{100 \times 5} = 10\sqrt{5}$ b $\sqrt[3]{15} = 15^{\frac{1}{3}}$

1 Calculate the following, leaving your answers in surd form.

a $\sqrt{5} \times \sqrt{6}$ b $\sqrt{7} \times \sqrt{11}$ c $\sqrt{10} \times \sqrt{13}$ d $\sqrt{2} \times \sqrt{17}$

e $\sqrt{3} \times \sqrt{19}$ f $\sqrt{7} \times \sqrt{31}$ g $\sqrt{13} \times \sqrt{11}$ h $\sqrt{7} \times \sqrt{23}$

i $\sqrt{17} \times \sqrt{13}$ j $\sqrt{11} \times \sqrt{23}$

2 Write these in their simplest form.

a $\sqrt{20}$ b $\sqrt{44}$ c $\sqrt{28}$ d $\sqrt{45}$ e $\sqrt{63}$

f $\sqrt{135}$ g $\sqrt{80}$ h $\sqrt{112}$ i $\sqrt{150}$ j $\sqrt{250}$

3 Calculate the following leaving your answers in surd form.

a $3\sqrt{2} \times 7\sqrt{3}$ b $6\sqrt{5} \times 3\sqrt{7}$ c $2\sqrt{11} \times 5\sqrt{3}$

d $7\sqrt{6} \times 11\sqrt{5}$ e $15\sqrt{5} \times 6\sqrt{2}$ f $2\sqrt{13} \times 7\sqrt{5}$

g $5\sqrt{11} \times 3\sqrt{7}$ h $3\sqrt{15} \times 4\sqrt{13}$ i $5\sqrt{14} \times 2\sqrt{5}$

j $7\sqrt{17} \times 3\sqrt{5}$

4 Write these using index notation.

a $\sqrt{10}$ b $\sqrt{50}$ c $\sqrt{90}$ d $\sqrt[3]{20}$ e $\sqrt[3]{100}$ f $\sqrt[3]{10}$

5 Work out these.

a $64^{\frac{1}{2}}$ b $144^{\frac{1}{2}}$ c $81^{\frac{1}{2}}$ d $64^{\frac{1}{3}}$

e $125^{\frac{1}{3}}$ f $1000^{\frac{1}{3}}$ g $512^{\frac{1}{3}}$ h $343^{\frac{1}{3}}$

6 Calculate these leaving your answer in index form.

a $5^{\frac{1}{3}} \times 5^2$ b $3^{\frac{1}{2}} \times 3^3$ c $6^{\frac{1}{2}} \times 6^5$ d $7^{\frac{1}{3}} \times 7^4$

e $10^{\frac{1}{3}} \times 10^3$ f $12^{\frac{1}{3}} \times 12$ g $15^{\frac{1}{2}} \times 15$ h $11^{\frac{1}{2}} \times 11^0$

7 Find

i the area

ii the diagonal length of the rectangle.

$(170\sqrt{2})$ mm

$(14\sqrt{50})$ mm

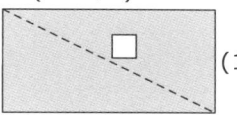

Example

a Write 2.5×10^4 as an ordinary number.

b Write $396\,000$ in standard form.

- -

a $2.5 \times 10^4 = 25\,000$ (The decimal point has to be moved 4 places to the right.)

b $396\,000 = 3.96 \times 10^5$ (The decimal point must go between the first two digits when the number is expressed in standard form, but its place in the ordinary number is 5 places to the right of that.)

1 Write these as ordinary numbers.

a	3×10^4	**b**	4×10^6	**c**	8.2×10^2	**d**	9.1×10^5
e	4.73×10^4	**f**	6.94×10^3	**g**	1.356×10^5	**h**	4.229×10^7
i	6.231×10^2	**j**	3.4581×10^3				

2 Write these in standard form.

a	$700\,000$	**b**	3000	**c**	$76\,000$	**d**	520
e	$3\,930\,000$	**f**	$15\,600$	**g**	$223\,100$	**h**	$86\,140\,000$
i	749.6	**j**	8536.1	**k**	39.68	**l**	42

3 Write these in standard form.

a The distance from the Earth to the Moon is $390\,000$ km.

b The distance from the Earth to the Sun is $150\,000\,000$ km.

c The circumference of the Earth is $40\,000$ km.

d The speed of sound is 1200 km/h.

e One gram of hydrogen occupies a volume of $11\,200$ cm³.

f The distance from the Sun to the planet Pluto is $5\,920\,000\,000$ km.

4 Write these in standard form.

a	28×10^3	**b**	43×10^5	**c**	16.4×10^4	**d**	39.2×10^6
e	351×10^2	**f**	911×10^4	**g**	0.52×10^7	**h**	0.67×10^6

5 A swimming pool has length 1.95×10^3 cm and width 9.5×10^2 cm. Find its perimeter in **a** centimetres **b** metres.

MyMaths.co.uk

1051   SEARCH

Example

a Express 2.7×10^{-2} as a decimal.

b Express 0.00021 in standard form.

- -

a $2.7 \times 10^{-2} = 0.027$ (The decimal point has to be moved 2 places to the left.)

b $0.00021 = 2.1 \times 10^{-4}$ (The decimal point must go between the first two digits when the number is expressed in standard form, but its place in the ordinary number is 4 places to the left of that.)

1 Work out these.

a	$531 \div 10^2$	**b**	$6500 \div 10^3$	**c**	$24000 \div 10^5$	**d**	$12 \div 10^3$
e	$26 \div 10^4$	**f**	55×10^{-2}	**g**	81.6×10^{-3}	**h**	3.2×10^{-4}
i	5.6×10^{-5}	**j**	0.54×10^{-3}				

2 Express these as a decimal.

a	2×10^{-3}	**b**	5×10^{-5}	**c**	3.6×10^{-2}	**d**	4.2×10^{-4}
e	6.32×10^{-6}	**f**	9.15×10^{-3}	**g**	6.256×10^{-5}	**h**	9.857×10^{-2}
i	6.352×10^{-1}	**j**	1.001×10^{-4}				

3 Express these in standard form.

a	0.0005	**b**	0.003	**c**	0.000081	**d**	0.0000045
e	0.000594	**f**	0.00776	**g**	0.01346	**h**	0.006051
i	0.6727	**j**	0.99	**k**	0.3		

4 Express these in standard form.

a Ultrasonic waves of wavelength $0.0095\,m$ are used in depth sounding.

b Light waves of wavelength $0.00006\,cm$ are emitted from yellow street lights.

c A salt grain, which is cubic in shape, has a side length of $0.004\,cm$.

d A grain of sand has a volume of $0.00015\,mm^3$.

5 A piece of adhesive tape of thickness $5 \times 10^{-3}\,cm$ is stuck onto a piece of paper of thickness $1.0 \times 10^{-2}\,cm$. Find the combined thickness.

Example

Construct the triangle ABC from the information shown. Measure the length of the third side and the size of the other two angles. (Use a ruler and a protractor.)

BC = 5 cm, angle B = 37°, angle C = 53°.

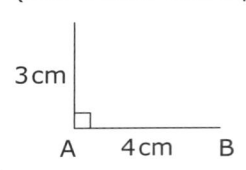

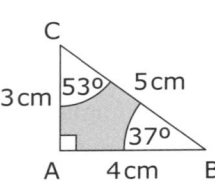

For all questions a protractor and a ruler are required.

1 For each of these construct the triangle from the starting information shown. Measure the length of the third side and the size of the other two angles.

a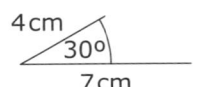

b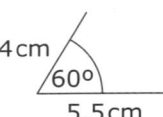

c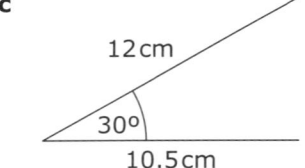

d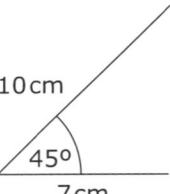

2 For each of these construct the triangle from the starting information shown. Measure the lengths of the other two sides.

a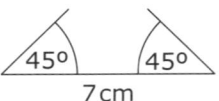

b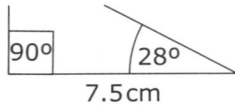

c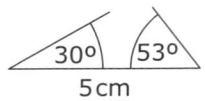

d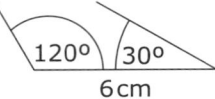

MyMaths.co.uk

Q 1090   SEARCH

Construct the triangle ABC with AB = 4 cm and
AC = BC = 2.5 cm. Measure the three angles in the triangle.

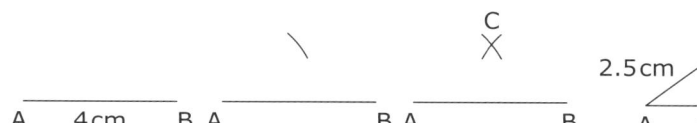

Measurement with a protractor shows that angle A = angle B = 37° and
that angle C = 106°.

1 Construct these triangles. Measure the three angles of each triangle.

a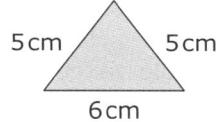

b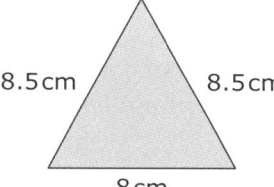

c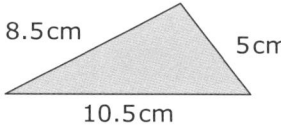

d

2 Construct these right-angled triangles. Measure
  i the missing side length   **ii** the missing angles.

a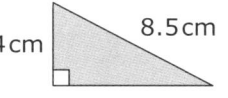

b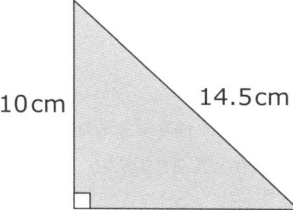

c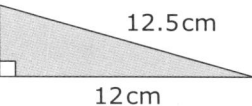

d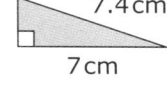

Construct the locus of the points that are equidistant from AB and BC.
(Or the bisector of angle ABC.)

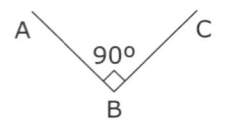

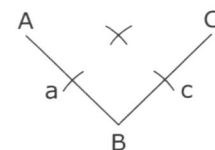

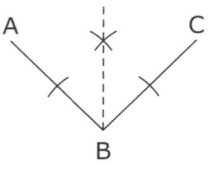

1   Copy the lines illustrated and bisect them. (The line bisector is the locus of the points that are equidistant from the two ends.)

   a _____   **b** _____   **c** _____
             8 cm   5 cm   3.6 cm

2   Copy the angles illustrated and bisect them. (The angle bisector is the locus of the points that are equidistant from the two lines.)

   a 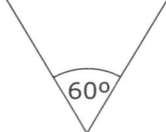   **b**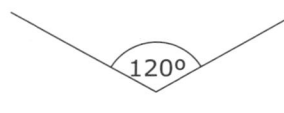

   c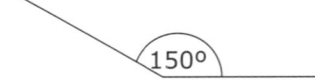

3   Copy the line AB. Draw on your copy the locus of the points that are

   i   8 cm from A

   ii 6 cm from B.

   A   10 cm   B

   iii Mark on the two points where both conditions apply and label them X and Y.

   iv Join AX, XB, BY and YA. What kind of figure have you drawn?

MyMaths.co.uk

Q 1089, 1147   SEARCH

Find the unknown sides.

a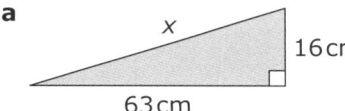

b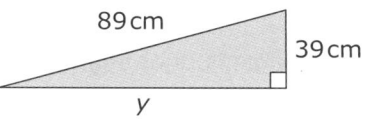

---

a $x^2 = 63^2 + 16^2 = 3969 + 256 = 4225$, therefore $x = \sqrt{4225} = 65$ cm

b $y^2 = 89^2 - 39^2 = 7921 - 1521 = 6400$, therefore $y = \sqrt{6400} = 80$ cm

1 Find the unknown area of the third square.

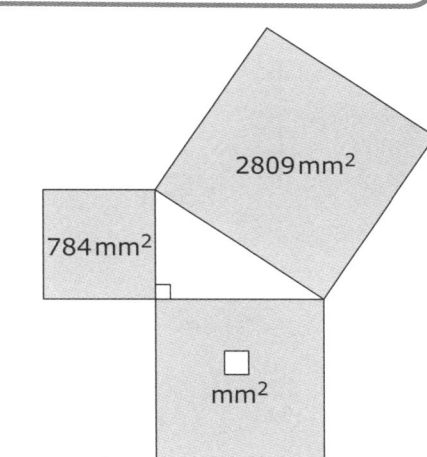

2809 mm²

784 mm²

☐ mm²

2 For these find the missing side.

a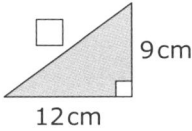

b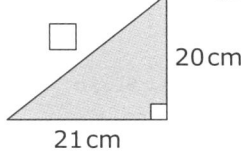

c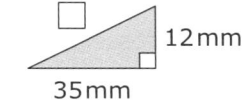

d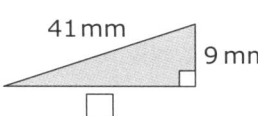

3 For these find the missing sides.

a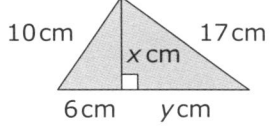

b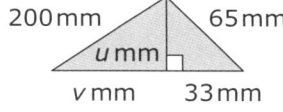

12e Pythagoras' theorem 2

Example

Use Pythagoras' theorem to decide if the triangle is right angled.

- -

If the triangle is right angled 4^2 and $2.6^2 + 3^2$ would have to be equal, but $4^2 = 16$ and $2.6^2 + 3^2 = 6.76 + 9 = 15.76$. Therefore the triangle is not exactly right angled.

3 cm   4 cm

2.6 cm

1 Use Pythagoras' theorem to decide if each of these triangles is right angled or not.

a
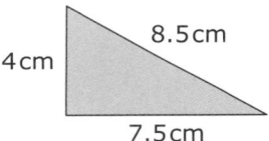

b
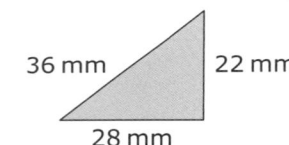

c 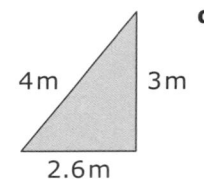 **d**
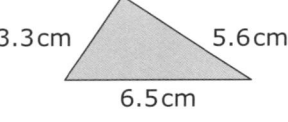

2 a A wire stay is supporting a radio mast.
   Find the length of the stay.
b Find the width of the gate.

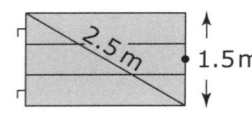
40 m   42 m   2.5 m   1.5 m

3 Find
   i the vertical height   **ii** the area for each of these isosceles triangles.

a
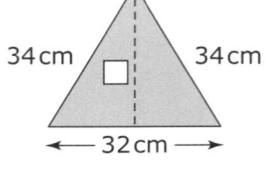

b
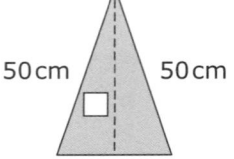

c
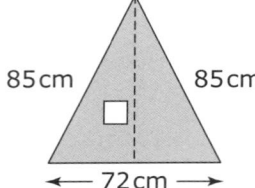

MyMaths.co.uk   Q 1112   **SEARCH**

13a Sequences and terms

Write the first five terms of the sequence with the term-to-term rule 'start with 900, divide by 3 and add 60'.

The first five terms are 900, 360, 180, 120 and 100.

1   Write the first five terms of the sequences with these term-to-term rules.
   a   Start with 3, double and subtract 2.
   b   Start with 4, double and subtract 3.
   c   Start with 1200, halve and subtract 40.
   d   Start with 1000, halve and add 60.
   e   Start with $\frac{1}{2}$, double and add 3.
   f   Start with $\frac{1}{4}$, multiply by 4 and add 2.

2   Two sequences both start with a first term of 6.
    Use these two function machines to find the first five terms for each sequence.

   a   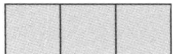   b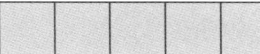

3   Look at the patterns below. Draw the next three patterns for this sequence.

4   State the term-to-term rule and give the next three terms for each of these sequences.
   a   5, 8, 11, 14, ...   b   33, 28, 23, 18, ...   c   1, 5, 13, 29, ...
   d   5, 7, 11, 19, ...   e   2, 4, 10, 28, ...   f   1, 2, 6, 22, ...
   g   192, 96, 48, 24, ...   h $\frac{1}{2}$, 2, 5, 11, ...   i $\frac{1}{2}$, 1, 4, 22, ...
   j   500 000, 100 000, 20 000, 4000, ...

Find **a** the position-to-term rule

b the 50th term for the sequence −1, 1, 3, 5, …

a The sequence grows by adding 2 to each term. The first four terms of the two times table are 2, 4, 6 and 8 and the sequence terms are less than these by 3. Therefore the position-to-term rule is 'multiply the position number by 2, then subtract 3'.

b The 50th term is 2 × 50 − 3 = 100 − 3 = 97.

1 For each of these sequences find

i the position-to-term rule

ii the next two terms

iii the 20th term.

a 4, 7, 10, 13, 16, …   **b** 3, 7, 11, 15, 19, …

c 5, 7, 9, 11, 13, …   **d** 2, 7, 12, 17, 22, …

e −1, 4, 9, 14, 19, …   **f** 48, 46, 44, 42, 40, …

g 56, 52, 48, 44, 40, …   **h** $5\frac{1}{2}$, 6, $6\frac{1}{2}$, 7, $7\frac{1}{2}$, …

i $-1\frac{1}{2}$, −1, $-\frac{1}{2}$, 0, $\frac{1}{2}$, …   **j** $5\frac{1}{2}$, 5, $4\frac{1}{2}$, 4, $3\frac{1}{2}$, …

2 For each of these sequences find

i the position-to-term rule

ii the missing terms

iii the 20th and 50th terms.

a 1, 4 … 10 …   **b** 7, 12 … 22 …   **c** 1 … 9 … 17

d 98 … 94 … 90   **e** 47 … 41, 38 …   **f** 35 … 25, 20 …

3 For each sequence

i draw the next two patterns

ii find the position-to-term rule for the number of tiles required

iii find the number of tiles required for the 10th pattern.

a      **b**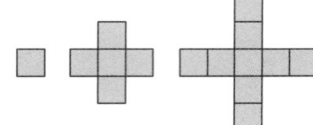

Example

Find the general term for the sequence given in the worked example for Homework 13b.

The position-to-term rule is 'multiply the position number by 2, then subtract 3'. Therefore the general term, $T(n)$ is given by $T(n) = 2n - 3$, where n is the position number.

1 For each of these sequences find
  i the position-to-term rule
  ii an expression for the nth (or general) term
  iii the 10th and 50th terms.

  a 7, 9, 11, 13, 15, ...　　　**b** 6, 9, 12, 15, 18, ...
  c 6, 10, 14, 18, 22, ...　　 **d** 6, 11, 16, 21, 26, ...
  e 2, 7, 12, 17, 22, ...　　　**f** −1, 3, 7, 11, 15, ...
  g −3, −1, 1, 3, 5, ...　　　 **h** −3, 0, 3, 6, 9, ...
  i 22, 19, 16, 13, 10, ...

2 Find the first six terms of the sequences with these general terms.

  a $2n + 6$　　**b** $3n + 5$　　**c** $4n + 4$　　**d** $10n + 2$　　**e** $2n - 6$
  f $3n - 8$　　**g** $5n - 7$　　**h** $6n - 10$　　**i** $20 - 5n$　　**j** $25 - 7n$
  k $10 - 3n$　　**l** $40 - 8n$　　**m** $2(n + 4)$　　**n** $4(n - 2)$

3 For each of these sequences find
  i the missing terms
  ii the general term
  iii the 20th term.

  a 10 ... 14 ... 18　　　　 **b** 13, 16 ... 22 ...　　　 **c** ... −3, 0 ... 6
  d −6 ... 2, 6 ...　　　　　**e** ... 2 ... −4, −7 ...　　 **f** ... 2, −2, −6 ...

4 The number of dots in these patterns forms a sequence.
  a Write down the number of dots for each pattern.
  b Find the general term for the sequence.
  c Find the number of dots in the 12th pattern.

$n = 1$　　　　　　　　　2　　　　　　　　　　3

13d Real–life sequences

A car starts from rest, reaches a speed of 4 km/h after 1 s and its
speed increases by 5 km/h for every second afterwards for 6 s.

a   Copy and complete the sequence table.

Time in seconds (position number)	1	2	3	4	5	6
Speed in km/h (term)	4					

b   State the term-to-term rule.

c   Find the position-to-term rule.

- -

a

Time in seconds (position number)	1	2	3	4	5	6
Speed in km/h (term)	4	9	14	19	24	29

b   The term-to-term rule is 'start with 4 and add 5'.

c   The position-to-term rule is 'multiply the position number by 5
and subtract 1' or $5n - 1$.

---

1   One day it begins to rain very heavily where Suzanne lives and the
water in the rain barrel outside her house is 12 cm deep after
1 hour and the depth increases by 8 cm for every hour afterwards
for 6 hours.

a   Copy and complete the sequence table.

Time in hours (position number)	1	2	3	4	5	6
Depth of the water (cm) (term)	12					

b   State the term-to-term rule.

c   Find the position-to-term rule and the general term.

2   The height of a planted tree in metres after n years is given by
$T(n) = \frac{1}{2}n + 1$.

a   Write a sequence for the height of the tree after 1, 2, 3, 4, 5
and 6 years.

b   By how much does the tree grow each year?

a Write the first five terms in the sequence
$T(n + 1) = T(n) + 5$, $T(1) = 6$.

b Describe this linear sequence using a recursive formula.
10, 14, 18, 22, ...

- -

a 6   6 + 5 = 11   11 + 5 = 16   16 + 5 = 21   21 + 5 = 26
6, 11, 16, 21, 26

b $T(1) = 10$
The rule is *add 4 to the previous term*
$T(n + 1) = T(n) + 4$

1 Write the first five terms of each of these sequences.
- **a** $T(n + 1) = T(n) + 4$, $T(1) = 5$ **b** $T(n + 1) = T(n) + 6$, $T(1) = 0$
- **c** $T(n + 1) = T(n) - 3$, $T(1) = 5$ **d** $T(n + 1) = T(n) - 2$, $T(1) = -3$
- **e** $T(n + 1) = T(n) - 6$, $T(1) = 1$ **f** $T(n + 1) = 10 - T(n)$, $T(1) = 2$

2 Write the first five terms of each of these sequences.
- **a** $T(n + 1) = 3T(n)$, $T(1) = 1$ **b** $T(n + 1) = 5T(n) + 2$, $T(1) = 0$
- **c** $T(n + 1) = 4T(n) - 3$, $T(1) = 2$ **d** $T(n + 1) = 2T(n) - 1$, $T(1) = 6$

3 Describe each of these linear sequences using a recursive formula.
- **a** 3, 10, 17, 24, ...   **b** 4, 9, 14, 19, ...
- **c** 2, 11, 20, 29, ...   **d** 40, 36, 32, 28, ...
- **e** 10, 6, 2, −2, ...   **f** 5, −1, −7, −13, ...

4 Marnie starts up a savings account with £200 and saves £25 each month.
- **a** Continue the sequence to show how much money Marnie has in her account for the first six months.
200, 225, ...
- **b** Describe the sequence using a recursive formula.

5 Josh buys a sofa that costs £400. He pays £20 per month until he repays the full amount.
- **a** Continue the sequence to show how much Josh has left to pay.
400, 380, ...
- **b** Describe the sequence using a recursive formula.

The midpoints of four faces of the cuboid, A, B, C and D are joined to make a quadrilateral ABCD.
What kind of quadrilateral will be drawn?

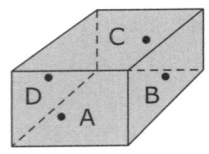

ABCD is a rhombus. It is easy to see that this is the case by viewing the rectangle from above.

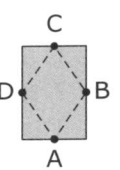

1 **a** On isometric paper draw as many different solids as possible by using
  i four cubes
  ii two prisms whose ends are isosceles, right-angled triangles
  iii two prisms whose ends are equilateral triangles.

i 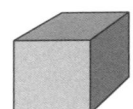   **ii**    **iii**

  b For each solid, state the number of faces (*f*), the number of vertices (*v*) and the number of edges (*e*).
  c Check if the formula $f + v = e + 2$ is true for each solid.

2 The pyramid illustrated is square based and the four sloping faces are equilateral triangles. Describe fully the properties of triangle APC (or BPD).

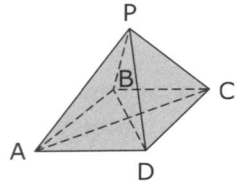

3 The pyramid illustrated has three faces which are right-angled isosceles triangles and all are congruent to each other. Describe fully the properties of
  a triangle APC   **b** triangle ABR (or CBR).

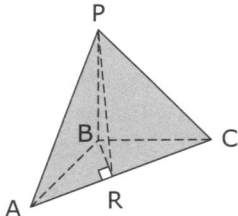

Example

Draw the front elevation, side elevation and plan view for the triangular prism illustrated.

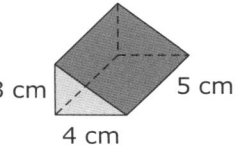

3 cm   5 cm
4 cm

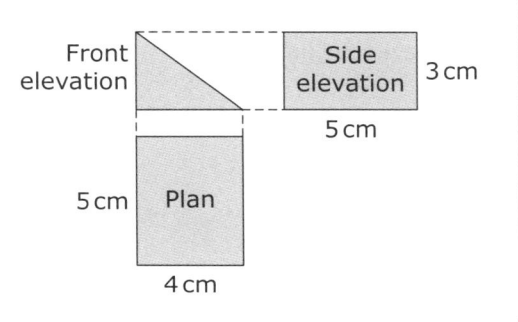

Front elevation

Side elevation   3 cm

5 cm

5 cm   Plan

4 cm

1 For these solids draw the front elevation, side elevation and plan view.

a

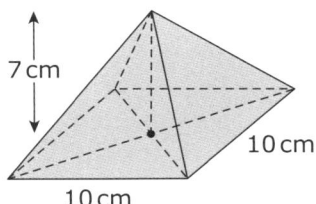

7 cm
6 cm
8 cm

b

5 cm
10 cm
8 cm

c

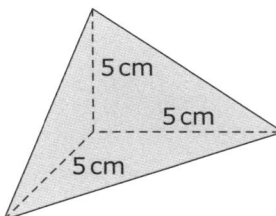

7 cm
8 cm
7 cm

d

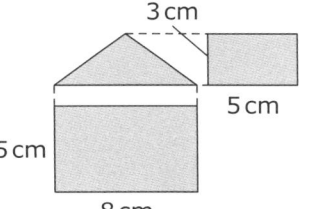

7 cm
10 cm
10 cm

e

5 cm
5 cm
5 cm

2 The front elevation, side elevation and plan view are drawn for these solids. Draw the solids on isometric paper.

a

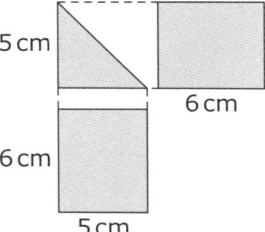

3 cm
5 cm
5 cm
8 cm

b

5 cm
6 cm
6 cm
5 cm

Example

Draw one plane of symmetry that a cylinder has.

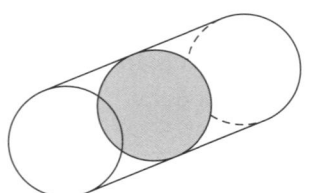

This is one of many planes of symmetry in a cylinder.

1 How many planes of symmetry does this cuboid have?

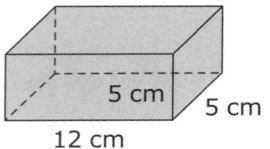

5 cm

5 cm

12 cm

2 Look at the pyramids.

For pyramid **a**, name the triangle that lies in a plane of symmetry.

For pyramid **b**, name two triangles that lie in a plane of symmetry.

a

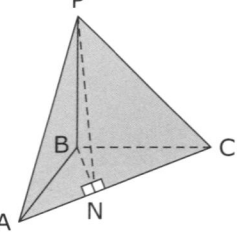

b

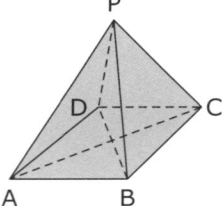

3 Look again at pyramid **b** in question **2**. How many planes of symmetry does this pyramid have?

Find the surface area of the triangular prism illustrated in the Example for Homework 14b, given that the sloping edges are of length 5 cm.

The area consists of two triangular faces and three rectangular ones.
Therefore the surface area

$$= \frac{1}{2} \times 4 \times 3 + \frac{1}{2} \times 4 \times 3 + 4 \times 5 + 3 \times 5 + 5 \times 5$$

$$= 6 + 6 + 20 + 15 + 25 = 72 \, cm^2$$

1 Find the surface area of these.

a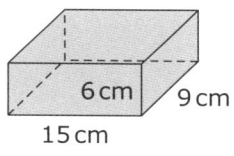
6 cm   9 cm   15 cm

b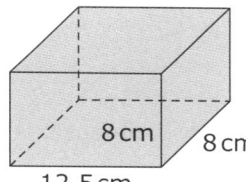
8 cm   8 cm   12.5 cm

c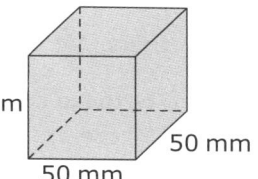
50 mm   50 mm   50 mm

2 Find the surface area of these.

a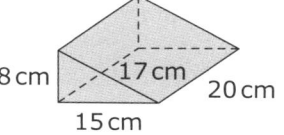
8 cm   17 cm   20 cm   15 cm

b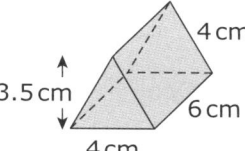
4 cm   3.5 cm   6 cm   4 cm

c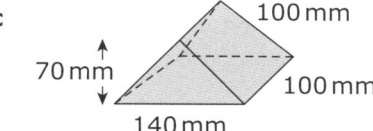
100 mm   70 mm   100 mm   140 mm

3 a The three different surface areas of a cuboid are 120, 90 and 48 cm². Find
   i the total surface area
   ii the length, width and height of the cuboid.

  b Repeat part **a** for a cuboid with three different surface areas equal to 96, 60 and 40 cm².

Example

Find the volume of the triangular prism in the Example for Homework 14b.

Volume = Area of triangular end × length

$$= \frac{1}{2} \times 3 \times 4 \times 5 = \frac{1}{2} \times 60 = 30 \, cm^3$$

1 Look at question **1** in Homework 14d. Find the volume of the cuboids **a**, **b** and **c**.

2 Look at question **2** in Homework 14d. Find the volume of the triangular prisms **a**, **b** and **c**.

3 For these cuboids find
  i the area of the face whose dimensions are given
  ii the missing dimension.

a
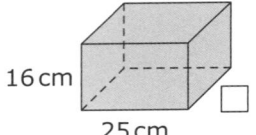

16 cm

25 cm

Volume = 8000 cm³

b
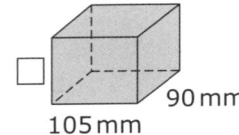

90 mm

105 mm

Volume = 756000 mm³

4 For these cuboids, which both have a square cross-section, find
  i the area of the square cross-section
  ii the side dimension of the square cross-section.

a
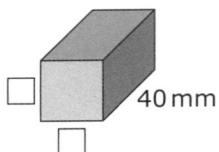

40 mm

Volume = 25000 mm³

b
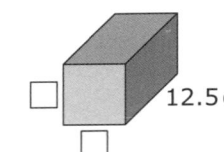

12.5 cm

Volume = 800 cm³

5 Find
  a the area of the circular end of the cylinder
  b the volume of the cylinder.

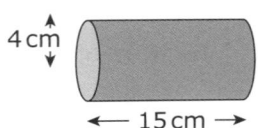

4 cm

← 15 cm →

Fizzpop is sold in packs that contain 4 or 6 cans. If a four-can pack costs £1.80, use direct proportion to find the cost of a six-can pack.

The cost of a six-can pack is $\frac{6}{4}$ × £1.80 = £2.70

1 The extension of a spring is directly proportional to the mass that is hung on its end. Copy and complete the table below which refers to a spring that Kanika experimented with.

Mass hanging (g)	50	60	70	80	90			
Extension produced (cm)	15					30	33	36

2 a George, Amid and Chen are standing together on a sunny day. Their heights are 165, 150 and 140 cm respectively. If George's shadow is 231 cm long, find the length of
  i Amid's shadow   **ii** Chen's shadow.
  b Jean, Rita and Barbara then arrive at the same place and their shadows have lengths of 224, 203 and 189 cm, respectively. Find the height of each of the three girls.

3 Afiya has a set of four similar set squares.

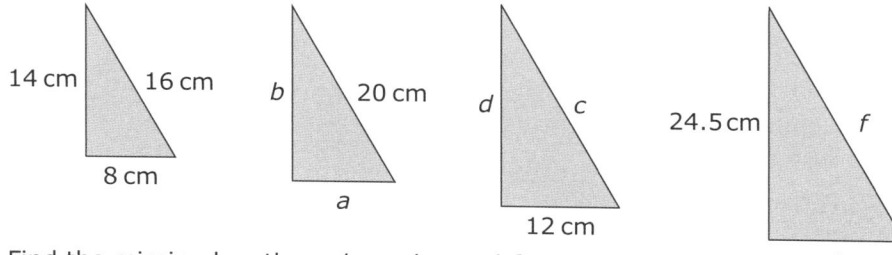

Find the missing lengths a, b, c, d, e and f.

4 Copy and complete the approximate conversion table for inches and centimetres.

Inches	10	15	20	25			
Centimetres	25				75	87.5	100

Example

A tennis club has 80 members and 44 of them are men.

Find the percentage who are **a** men **b** women.

- -

a The proportion who are men is $\frac{44}{80}$.

Therefore the percentage who are men = $\frac{44}{80} \times 100\% = 55\%$

b The proportion who are women is $(80 - 44) \div 80 = \frac{36}{80}$

Therefore the percentage who are women is $\frac{36}{80} \times 100\% = 45\%$

1 Copy and complete the table about Bob's examination results.

Subject	English	French	History	Geography	Craft	Maths	Science
Mark	26 out of 40	33 out of 60	48 out of 80	32 out of 50	42 out of 75	36 out of 90	54 out of 120
Fraction of total	$\frac{13}{20}$						
Percentage of total							

List his subject marks as percentages in order of merit.

2 Copy and complete the table which gives details of the pupils in each year at Holly Park School.

Year	7	8	9	10	11
Number of pupils	240	225	210	250	200
Number of boys	132	108	84		
Number of girls				110	124
Percentage who are boys					
Percentage who are girls					

Find

a the total number of pupils in the school

b the total number of boys

c the total number of girls

d the percentage of the total who are boys

e the percentage of the total who are girls.

15c Ratio

The angles x and y in the triangle
shown have sizes in the ratio of
$2:7$. Find the value of both x and y.

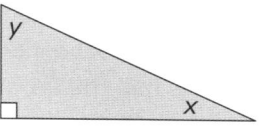

- -

$2 + 7 = 9$; the sum of x and y must be $90°$ and $90 \div 9 = 10$.
Therefore $x = 2 \times 10 = 20°$ and $y = 7 \times 10 = 70°$.

1 Write these ratios in their simplest form. (Divide both numbers
by their highest common factor.)

a	$8:12$	**b**	$12:16$	**c**	$12:15$	**d**	$9:12$
e	$15:50$	**f**	$25:30$	**g**	$24:30$	**h**	$21:36$

2 Write these ratios in their simplest form. (Divide all three numbers by
their highest common factor.)

a	$10:12:16$	**b**	$14:18:20$	**c**	$12:16:24$	**d**	$6:15:18$
e	$12:18:30$	**f**	$14:28:35$	**g**	$24:40:48$	**h**	$18:45:54$

3 Write these ratios in their simplest form by first changing the quantities
into the same units.

a	$60\text{p}:£2$	**b**	$25\,\text{mm}:4\,\text{cm}$	**c**	$75\,\text{cm}:5\,\text{m}$
d	$625\,\text{g to }2\,\text{kg}$	**e**	$900\,\text{ml to }3\,\text{litres}$	**f**	$54\,\text{min to }4\,\text{hours}$
g	$45\,\text{s to }3\,\text{min}$	**h**	$1750\,\text{mm to }4\,\text{m}$		

4 Divide these quantities in the ratio given.

Quantity	£120	90 m	105 kg	225 cm	£240	36 m	132 kg	108 cm	105 g
Ratio	$3:7$	$7:8$	$2:5$	$4:5$	$5:7:8$	$2:3:4$	$2:3:6$	$3:4:5$	$3:5:7$

5 The lengths of the sides of an isosceles triangle are in the ratio of
$7:7:11$. Find the three side lengths if the perimeter of the triangle
is $200\,\text{mm}$.

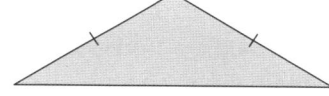

Example

At Southmead School 4 pupils out of 25 failed an English exam, whereas at Parkfield School 5 out of 31 failed the same exam. Which of the two schools had the lower proportion of failures?

At Southmead School the ratio was $4:25$ or $1:\left(\frac{25}{4}\right)$ which is $1:6.25$.

At Parkfield School the ratio was $5:31$ or $1:\left(\frac{31}{5}\right)$ which is $1:6.2$.

Therefore Parkfield School had the lower proportion of failures because 6.2 is less than 6.25.

1 Write each of these ratios in the form $1:n$.

a $4:7$	**b** $5:6$	**c** $8:11$	**d** $8:15$	**e** $5:7$
f $4:9$	**g** $5:13$	**h** $4:13$	**i** $5:18$	**j** $5:22$

2 a At Brown's the Builders the ratio of managers to workers is $4:37$, whereas at Carr's the Constructors the corresponding ratio is $10:93$. Which firm has the lower proportion of managers?

b At Carl's Cafe, orange squash is served from a machine that mixes orange syrup and water in the ratio $5:44$, whereas at Billy's Bar a similar machine mixes them in the ratio $4:35$. Which of the two serves the weaker flavoured squash?

3 a At Grange Park School the ratio of boys to girls is $10:11$. If there are 480 boys, find

    i the number of girls   **ii** the total number of pupils.

b A railway station has three platforms whose lengths are in the ratio $5:6:9$. If the shortest has a length of 75 m, find the lengths of the other two.

c One day the ticket office at the railway station in part **b** sold 540 tickets and the passengers paid by using cash, credit card or cheque in the ratio $10:5:3$. If 300 paid in cash, find

    i the number who paid by credit card

    ii the number who paid by cheque.

MyMaths.co.uk

Q 1103   SEARCH

Example

In Fadi's street the ratio of houses to bungalows is $11:4$.

What proportion of the dwellings are

a houses   **b** bungalows?

- -

$11 + 4 = 15$. Therefore,

a the proportion of houses is $\frac{11}{15}$

b the proportion of bungalows is $\frac{4}{15}$.

1 Copy and complete the table about the pupils at High Lane School.

Year	7	8	9	10	11	12	13
Ratio (boys to girls)	$4:5$	$6:7$	$8:7$	$13:12$	$9:10$	$11:12$	$15:13$
Proportion who are boys							
Proportion who are girls							

2 For each of these calculate the whole (100%).

  a 20% is £48   **b** 15% is £36   **c** 5% is 25 cm

  d 30% is 6 kg   **e** 20% is £3.50   **f** 35% is 2.8 m

  g 15% is 7.2 g   **h** 25% is £4.50   **i** 16% is £8.40

3 Copy and complete this table about a sale.

Article	Original price	Sale price	Reduction	Reduction (%)
DVD	£25	£21		
Bike	£250		£30	
CD Player		£36	£12	
Radio		£13.60	£2.40	
Shirt	£20			12%
Pullover			£1	8%
Tie			45p	10%
Pair of shoes		£5.40		20%

Example

Eva's car travels 80 km on 5 litres of petrol. Assuming that direct proportion applies, find

a the distance that her car can travel on 3 litres of petrol

b the number of litres her car will consume in travelling 100 km.

- -

a On 1 litre the car will travel 80 ÷ 5 = 16 km

Therefore on 3 litres it will travel 3 × 16 = 48 km

b For 1 km the car requires 5 ÷ 80 = 0.0625 litres

Therefore for 100 km the car requires 0.0625 × 100 = 6.25 litres

1 Jordan is doing an experiment with a coiled spring and finds that it is extended 4 cm when a mass of 200 g is hung on its end. Assuming that direct proportion applies, find

a the mass that will extend the spring by

    i 9 cm   **ii** 5 cm   **iii** 7 cm   **iv** 4.5 cm

b the distance that the spring will be extended by a hanging mass of

    i 300 g   **ii** 550 g   **iii** 375 g   **iv** 525 g.

2 One very wet day, the rain barrel outside Marlon's house began to fill up steadily and after 3 hours the water was 7.5 cm deep.

a Find the depth of the water after

    i 4 hours   **ii** 6 hours.

b Find the time after which the water was

    i 5 cm deep   **ii** 12.5 cm deep.

3 Martin often goes mountain climbing. One day he climbed Ben Nevis in Scotland and the table shows his altitude after certain times. Check that these figures are in direct proportion and state his altitude gain per hour.

Time (hours)	0	2	$2\frac{1}{2}$	3	4	$4\frac{1}{2}$	5
Altitude (m above sea level)	0	536	670	804	1072	1206	1340

MyMaths.co.uk

Q 1036   SEARCH

Matt's monthly salary is £1200.
He plans his monthly outgoings
in a table.

Outgoing	Amount budgeted £
Rent	600
Food shopping	120
Bills	60
Transport	90
Mobile phone	30
Other	150

What **proportion** of Matt's
salary is spent on
a   transport
b   rent and bills?
In each case, give your answer
as a fraction in its lowest terms.

a $\dfrac{90}{1200} = \dfrac{9}{120} = \dfrac{3}{40}$

b $600 + \dfrac{60}{1200} = \dfrac{660}{1200} = \dfrac{66}{120} = \dfrac{33}{60} = \dfrac{11}{20}$

Use the table in the example to answer the questions.

1   a   How much money does Matt have left over at the end of the
month.

b   Matt saves half of the money that he has left over each month.
How long will it take him to buy a new laptop that costs £399.99?

2   What **proportion** of Matt's salary is spent on
a   food shopping   b   mobile phone
c   other expenses   d   bills and transport?
In each case, give your answer as a fraction in its lowest terms.

3   Matt is thinking about moving to a new flat that costs £650 to rent
each month. The new flat is closer to Matt's job, so he will save 50%
on transport.
Matt will also save one third on his monthly bills.
Would you advise Matt to move?

Example

A railway company runs roughly twice as many passenger trains as freight trains. One morning Ayo watches some trains pass and this was the sequence that he saw.

Passenger   Passenger   Passenger   Passenger   Freight   Freight

This sequence may look very improbable, but can you explain why it is actually very likely?

Perhaps he started watching during the morning or evening rush hour and probably freight trains do not operate then.

1 Describe these events using probability words.
   a If I go to the Sahara Desert it will be raining when I get there.
   b If I go to anywhere in the tropical area I will find that day and night are of about equal length.
   c If I go to the North Pole in December I will arrive on a sunny day.
   d On my way to work tomorrow morning I will find that all of the six sets of traffic lights on the way are showing green.

2 Ed cycles to school and he was late on every morning during a certain week. He told his teacher that the level crossing gates had stopped him on each of the mornings and the railway company do admit that there is an evens chance of them being shut. Do you think, though, that his teacher should have believed him? Explain your answer.

3 Candace has an old car and she knows that there is only an evens chance that it will start first thing in the morning. One week, however, she found that this was the sequence.

Monday	Tuesday	Wednesday	Thursday	Friday	Saturday
No	No	No	Yes	Yes	Yes

This sequence may look very improbable, but can you explain why it might be very likely? Give a reason for your answer.

MyMaths.co.uk

Q 1209   SEARCH

A square spinning dice is spun and the event X is
'the score is a prime number', whereas the event Y
is 'the score is a square number'.
Are these events

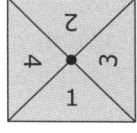

a   mutually exclusive
b   exhaustive?

a   The possible outcomes for event X are (2, 3) and the possible
    outcomes for event Y are (1, 4). Therefore the events are
    mutually exclusive because there is no possible outcome
    which is common to both.
b   The four possible outcomes are (1, 2, 3, 4). Therefore the
    events are exhaustive because every possible outcome is
    included in either X or Y.

1   The letters of the word EASY
    are drawn on cards as shown.
    The symmetry of each letter is
    indicated. If the cards are placed
    in a bag and one is drawn out at
    random, state whether or not these
    pairs of events are mutually exclusive.
    Explain your answers.

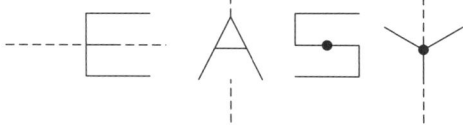

a   Event P is that the letter has rotational symmetry.
    Event Q is that the letter has line symmetry.
b   Event P is that the letter has rotational symmetry.
    Event R is that the letter is a vowel.
c   Are events P and R exhaustive? Explain your answer.

The letters of the name ETHEL are each printed on a card and the cards are placed in a bag. If a card is removed from the bag at random, what is the probability that it has the letter E written on it? Give the answer as

a a fraction in its simplest form **b** a decimal **c** a percentage.

- -

a Probability = number of cards with an E ÷ total number of cards = $\frac{2}{5}$

b $\frac{2}{5} = \frac{4}{10} = 0.4$

c $\frac{2}{5} = \frac{4}{10} = \frac{40}{100} = 40\%$

For each question below give all probability answers as a fraction (in its simplest form), a decimal and a percentage.

1 Simone has bought a large packet of sweets: 18 of them are orange flavoured, 15 of them are lemon flavoured, 12 of them are strawberry flavoured, 9 of them are raspberry flavoured and 6 of them are lime flavoured. Find the total number of sweets in her packet. If she picks a sweet out of the packet at random, what is the probability that its flavour will be

    a orange   **b** lemon   **c** strawberry

    d raspberry   **e** lime?

2 On a bus there are 15 men, 21 women, 9 boys and 3 girls. Find the total number of passengers on the bus. What is the probability that the first passenger to get off will be

    a a man   **b** a woman

    c a boy   **d** a girl?

Example

When possible, Rhema takes a shortcut through a park on her way to school and there is a gate at both the entrance and the exit to the park. The gates are never open because they are spring loaded, but she can find them either locked or unlocked. Draw
a a two-way table
b a tree diagram to show the possible outcomes.

a

1st gate		2nd gate	
		Locked (L)	Unlocked (U)
	Locked (L)	LL	LU
	Unlocked (U)	UL	UU

b

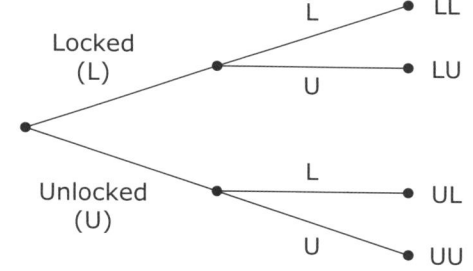

1 A football team can, of course, either win, draw or lose a match. Draw
   a a two-way table
   b a tree diagram to show all of the possible outcomes over two matches.

2 On her way to work Laura passes two sets of traffic lights either of which might be showing any of red, amber or green. Draw
   a a two-way table
   b a tree diagram to show all of the possible outcomes.
   c Which of the two do you think is the more convenient?

3 Josiah cycles to school and he passes three level crossings on the way which may be either open or closed. Draw a tree diagram to show the possible outcomes.

MyMaths.co.uk

Example

Bob played for a football team for a number of seasons and during that time he took 48 penalty kicks. He scored from 36 of them. Find from these figures the probability that he would score from a penalty kick. Give your answer as

a a fraction in its lowest terms   **b** a decimal   **c** a percentage.

- -

a The probability that he would score $= \frac{36}{48} = \frac{3}{4}$

b $\frac{3}{4} = \frac{75}{100} = 0.75$ **c** $0.75 = 75\%$

In all these questions give probability answers as a fraction (in its lowest form), a decimal and a percentage.

1 Laura, Beatriz, Deana, Jamuna and Simone recorded how many times they had found the traffic lights near to their workplace showing green. Here are their results.

	Laura	Beatriz	Deana	Jamuna	Simone
Number of journeys	60	50	40	20	30
Number of times green was displayed	15	12	8	6	9

a Find the probability of the lights showing green from each of the five results shown.

b Find the probability of the lights showing green from all the results taken together.

2 Jake plays football and he is the leading scorer for his team. One season he played in all of his club's matches and here is his goal scoring record.

Number of goals per match	0	1	2	3	4 or more
Number of matches	15	20	10	5	0

From these details find the probability that in any given match he will score

a no goals   **b** 1 goal   **c** 2 goals   **d** 3 goals.

MyMaths.co.uk

Q 1211, 1264   SEARCH

Marlon spun a tetrahedral dice 50 times and found that the score was four on 14 of these times. Find the probablity that the score is four from these figures. Express your answer as a decimal. Does your answer agree with the theoretical probability?

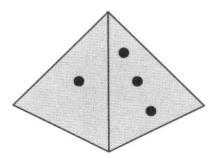

From Marlon's results the probability $= \frac{14}{50} = \frac{7}{25}$ which is equal to $\frac{28}{100}$ or 0.28.

Yes, because the theoretical probability is $\frac{1}{4}$ which is $\frac{25}{100}$ or 0.25. The two probabilities are therefore very close.

For the following question give probability answers as a fraction (in its lowest terms) and as a decimal.

1 Zoey has a gaming machine in her cafe, but she insists that it is for her customers' amusement only and not for making a profit. The machine is operated by 20p coins and she says that she has geared the outcomes to have these probabilities.

Outcome	Win 50p	Win 20p	Coin returned	Lose
Probability	$\frac{1}{10}$	$\frac{1}{5}$	$\frac{1}{4}$	$\frac{9}{20}$

One day, however, 120 people used the machine and the outcomes are shown on the table. Copy and complete the table.

Outcome	Win 50p	Win 20p	Coin returned	Lose
Number of people	12	15	18	75
Experimental probability				

Do the theoretical and experimental probabilities agree? Explain your answer.

16g Venn diagrams

A = {3, 4, 5, 6, 7}, B = {3, 4, 7} and C = {1, 2, 8, 9}.

Decide if the following sets are mutually exclusive or proper subsets.

a　A and B

b　A and C

- -

a　Every element in set B is also in set A.
　　B is a proper subset of A, B ⊂ A.

b　Set A and set C have no common elements.
　　A and C are mutually exclusive.

1　A = {1, 3, 5, 7, 9}, B = {2, 4, 6, 8}, C = {3, 9}
　　and D = {3, 4, 6, 9}.

　　Decide if the following sets are mutually exclusive,
　　proper subsets or neither.

　　a　A and B　　**b**　A and C　　**c**　A and D
　　d　B and C　　**e**　B and D　　**f**　C and D

2　Julie asks 40 people at her office if they
　　travelled to work by car or by train last week.
　　Julie used the sets C = {car} and T = {train}.
　　The results are shown on the Venn diagram.

　　a　On a copy of the diagram, shade the
　　　　region P(C' ∩ T).

　　b　Julie selects a person at random. Find P(C' ∩ T).

　　c　How many people didn't use a car or a train to travel to work
　　　　last week?

C　　T

20　5　8

7

3　Hari owns a bakery. He sorts the orders of
　　50 customers into C = {cookie} and
　　M = {muffin}.
　　A customer is chosen at random, find

　　a　the probability that they bought a
　　　　cookie or a muffin, but not both

　　b　P(M ∪ C')

　　c　P(M' ∪ C').

C　　M

15　10　20

5

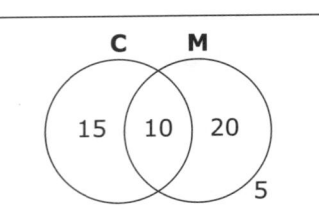

accuracy

You can round numbers to any degree of accuracy, for example, to the nearest power of 10 or to a given number of decimal places.

adjacent

Adjacent means 'next to'.

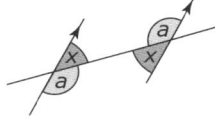

alternate angles

A pair of alternate angles are formed when a line crosses a pair of parallel lines. Alternate angles are equal.

angles in a triangle

Angles in a triangle add up to 180°.

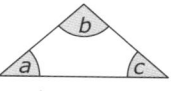

$a + b + c = 180°$

angles on a straight line

Angles on a straight line add up to 180°.

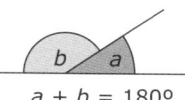

$a + b = 180°$

approximate, approximation

An approximate value is a value that is close to the actual value.

approximately equal to (≈)

Approximately equal to means almost the same size.

arc

An arc is a part of a curve.

area

The area of a surface is a measure of the space it covers. The units for area are square millimetre, square centimetre, square metre, square kilometre and so on.

average

An average is a representative value of a set of data.

axis, axes

An axis is one of the lines used to locate a point in a coordinate system.

| **bar chart** | A bar chart is a diagram that uses rectangles of equal width to display data. The frequency is given by the height of the rectangle. |

bar chart — A bar chart is a diagram that uses rectangles of equal width to display data. The frequency is given by the height of the rectangle.

bar-line chart — A bar-line chart is a diagram that uses lines to display data. The lengths of the lines are proportional to the frequencies.

base — In index notation, the base is the number which is to be raised to a power.

For example, in 5^3, 5 is the base.

base (of plane shape or solid) — The lower horizontal edge of a plane shape is called the base. Similarly, the base of a solid is its bottom face.

base

bias — An experiment is biased if not all outcomes are equally likely. A selection is biased if the members of the population do not have equal chances of being chosen.

bisect, bisector — A line that divides an angle or another line in half.

calculate, calculation — Work out using a mathematical procedure.

cancel, cancellation — A fraction is cancelled down by dividing the numerator and denominator by a common factor.

For example, $\dfrac{24}{40} \overset{\div 8}{\underset{\div 8}{=}} \dfrac{3}{5}$

capacity: litre — A measure of the amount of liquid a 3D shape will hold.

centre (of a circle) — The centre of a circle is the point in the middle.

centre

centre of enlargement — The centre of enlargement is the point from which an enlargement is measured.

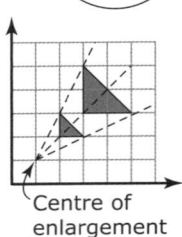

Centre of enlargement

centre of rotation	The centre of rotation is the fixed point about which a rotation takes place. 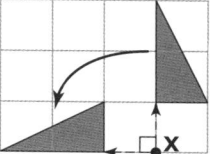
certain	An event that is certain will definitely happen.
chord	A chord is a line joining two points on the circumference of a circle. 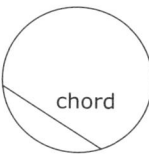
circle	A circle is a set of points that are the same distance from a fixed point, the centre.
circumference	The circumference is the distance around the edge of a circle. You calculate it using the formula: $C = 2\pi r$ where r is the radius. 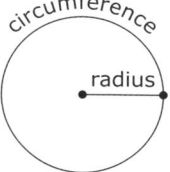
class interval	A class interval is a group that you put data into to make it easier to handle.
common denominator	A common denominator is an integer that is exactly divisible by all the denominators in a set of fractions, for example, the common denominator of $\frac{2}{3}$ and $\frac{1}{2}$ is 6.
common factor	A factor is a factor of two or more numbers. For example, 2 is a common factor of 4 and 10.
congruent, congruence	Congruent shapes are exactly the same shape and size.
construct	To form an equation from given facts or to draw a line, angle or shape accurately.
construction lines	Arcs and lines drawn during the construction of angles and lines.
continuous (data)	Continuous data can take any value between given limits, for example, less than 1 m.

coordinate pair	A coordinate pair is a pair of numbers that give the position of a point on a coordinate grid.
	For example, (3, 2) means 3 units across and 2 units up.
correlation	A measure of the relationship between two variables.
corresponding angles	A pair of corresponding angles is formed when a straight line crosses a pair of parallel lines. 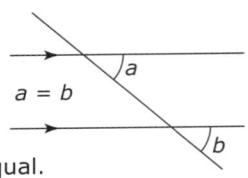 Corresponding angles are equal. $a = b$
corresponding sides	Corresponding sides in congruent shapes are equal in length.
cosine	The cosine of an angle in a right-angled triangle is the ratio of the adjacent side a, to the hypotenuse h. 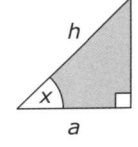 $$\cos x = \frac{a}{h}$$
cross-section	The cross-section of a solid is the 2D shape you get when you slice the solid perpendicular to its length.
cube, cube number	A cube number is the product of three equal integers, for example, $27 = 3 \times 3 \times 3$ so 27 is a cube number.
cube root	The value that gives the number when multiplied by itself twice.
	For example, $2 \times 2 \times 2 = 8$, so 2 is the cube root of 8, or $2 = \sqrt[3]{8}$.
cubed	A number multiplied by itself twice.
	For example, 2 cubed, written 2^3, is $2 \times 2 \times 2 = 8$.
data	Data are pieces of information.
data collection sheet	A data collection sheet is used to collect data. It is sometimes a list of questions with tick boxes.
decimal place (dp)	Each column after the decimal point is called a decimal place.
	For example, 0.65 has two decimal places (2 dp).

degree (°)	A measure of turn. There are 360° in a full turn.
denominator	The denominator is the bottom number in a fraction. It shows how many parts there are in total.
diagonal	A diagonal of a polygon is a line joining two vertices but not forming a side.

This is a diagonal.

diameter	A diameter is the distance across a circle through the centre.

diameter

difference	The difference between the terms of a linear sequence is constant (always the same).
digit	A digit is any of the numbers 0, 1, 2, 3, 4, 5, 6, 7, 8, 9.
dimensions	Flat 2D shapes have two dimensions: length and width or base and height

height

base

Solid 3D shapes have three dimensions:

length, width and height.

height

width

length

direct proportion	Two quantities are in direct proportion if they are always in the same ratio.
discrete (data)	Discrete data can only take certain definite values, for example, shoe sizes.
distance–time graph	A graph showing distance on the vertical axis and time on the horizontal axis.
distribution	Distribution describes the way data is spread out.
dividend, divisor	In this division sum: $6.25 \div 5 = 1.25$

this is the dividend	this is the divisor	this is the quotient

elevation	An elevation is an accurate drawing of the side or front view of a solid.
enlarge, enlargement	An enlargement is a transformation that multiplies all the sides of a shape by the same scale factor.
equally likely	Events are equally likely if they have the same probability.
equation	An equation is a statement linking two expressions that are equal in value.
equidistant	The same distance from a point or line.
equivalent, equivalence	Equivalent fractions are fractions with the same value, for example, $\frac{12}{20} = \frac{3}{5}$.
estimate	An estimate is an approximate answer.
evaluate	Find the value of an expression.
event	An activity or the result of an activity.
exact, exactly	Exact means completely accurate. For example, three divides into six exactly.
expand	You expand brackets by multiplying them out, for example, $3(2x - 5) = 6x - 15$.
expected frequency	The expected frequency of an event is the number of times it is expected to occur. Expected frequency = probability × number of trials.
experiment	An experiment is a test or investigation to gather evidence for or against a theory.
experimental probability	Experimental probability is calculated from the results of an experiment.
expression	An expression is a collection of numbers and symbols linked by operations that does not include an equals sign.
exterior angle	An exterior angle is made by extending one side of a shape.

face	A face is a flat surface of a solid.
factor	A number that divides exactly into another number. For example, 3 and 7 are factors of 21.
factorise	Writing a number or expression as a product of its factors. For example, $4a + 6 = 2(2a + 3)$.
fair	In a fair experiment there is no bias towards any particular outcome.
favourable outcome	A favourable outcome is a successful result of doing something. For example, throwing a 'six' with a fair dice.
formula, formulae	A formula is a statement that links variables.
fraction	A fraction is a way of describing a part of a whole. For example, $\frac{2}{5}$ of the shape shown is shaded.
frequency	Frequency is the number of times something occurs.
frequency diagram	A frequency diagram uses bars to display grouped data. The height of each bar gives the frequency of the group, and there is no space between the bars.
frequency table	A frequency table shows how often each event or quantity occurs.
function	A function is a rule. For example, $+ 2$, $- 3$, $\times 4$ and $\div 5$ are all functions.
function machine	A function machine links an input value to an output value by performing the function.
general term	The expression which relates its value to its position in the sequence.
gradient, steepness	A measure of the steepness of a line.
graph	A graph is a diagram that shows a relationship between variables.

| **greater than or equal to (≥)** | Greater than or equal to means equal to or more than.

For example, $x \geq 3$ means x can have any value from 3 upwards. |

grouped data

When there is a lot of data it is often easier to collect it into groups to see trends more easily. The end of one group must not overlap with the start of the next group:

$$0 < x \leq 10 \qquad\qquad 10 < x \leq 20$$

This group includes 10.   This group doesn't.

hectare

A hectare is a unit of area equal to $10\,000$ m^2.

highest common factor (HCF)

The HCF is the largest factor that is common to two or more numbers.

For example, the HCF of 12 and 8 is 4.

horizontal

Horizontal means level with the flat ground.

hypotenuse

The hypotenuse is the longest side in a right-angled triangle. It is opposite the right angle.

hypotenuse

hypothesis, hypotheses

A hypothesis is an unproved theory.

identity, identically equal to (≡)

The expressions in an identity are always equal.

For example, $3(x + 2) \equiv 3x + 6$ for all values of x.

image

The position of an object following a transformation.

impossible

An event is impossible if it cannot happen.

improper fraction

The numerator is greater than the denominator.

For example, $\frac{8}{5}$ is an improper fraction.

independent events

Two events are independent when one does not affect the outcome of the other.

For example, the outcome from flipping a coin has no effect on the outcome from rolling a dice.

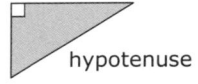

index, indices	The index of a number tells you how many of the number must be multiplied together. When a number is written in index notation, the index or power is the raised number.
	For example, the index of 4^2 is 2. The plural of index is indices.
index form	A number is in index form when it is written as a power: 5^3 is in index form.
index laws	The rules that tell you how to multiply and divide powers of the same base:
	$a^m \times a^n = a^{m+n} \quad a^m \div a^n = a^{m-n}$
index notation	A number is written in index notation when it is expressed as a power of another number.
	For example, 9 in index notation is 3^2.
integer	An integer is a positive or negative whole number (including zero).
intercept	The intercept is the length between the origin and the point where a line crosses the axis.
interior angle	An interior angle is inside a shape, between two adjacent sides.
interpret, interpretation	You interpret data whenever you make sense of it. What you write is your interpretation of the data.
intersect, intersection	Two lines intersect at the point where they cross.

intersection

interval	The size of a class or group in a frequency table.
inverse	An inverse operation has the opposite effect to the original operation.
inverse function	An inverse function undoes the effect of a function.
	For example the function $x \rightarrow 2x - 1$, maps 3 onto 5 and the inverse function
	$x \rightarrow \frac{x + 1}{2}$ maps 5 onto 3.

length: millimetre, centimetre, metre, kilometre	Length is a measure of distance. It is often used to describe one dimension of a shape.
less than or equal to (≤)	Less than or equal to means equal to or smaller than. For example, $x \leq 2$ means that x can have any value from 2 downwards.
line symmetry	A shape has line symmetry if it has a line of symmetry.

line of symmetry →

linear equation, linear expression, linear function, linear relationship	An equation, expression, function or relationship is linear if the highest power of any variable it contains is 1. For example, $y = 3x - 4$ is a linear equation, and its graph is a straight line.
linear sequence	The terms of a linear sequence increase by the same amount each time.
locus, loci	A locus is the position of a set of points, usually a line, that satisfies some given condition. Loci is the plural of locus.
lowest common multiple (LCM)	The LCM is the smallest multiple that is common to two or more numbers. For example, the LCM of 4 and 6 is 12.
mapping	A rule that can be applied to a set of numbers to give another set of numbers.
mean	The mean is an average value found by adding all the data and dividing by the number of pieces of data.
median	The median is an average found by taking the middle value when the data is arranged in size order.
mirror line	A mirror line is a line or axis of symmetry.

mirror line

modal class	The class with the highest frequency.
mode	The data value that occurs most often.
multiply out (expressions)	To multiply out a bracket you multiply each term inside by the term outside. For example, $3(x + 1)$ multiplied out is $3x + 3$.
mutually exclusive	Events that cannot both occur in one trial. For example, if you toss a coin once, you cannot get a Head and a Tail.
negative	A negative number is a number less than zero.
net	A net is a 2D arrangement that can be folded to form a solid shape.
numerator	The numerator is the top number in a fraction. It shows how many parts you are dealing with.
object	The object is the original shape before a transformation.
operation	A rule for processing numbers or objects. The arithmetic operations are addition, subtraction, multiplication and division.
opposite (sides, angles)	Opposite means across from. This side is opposite the shaded angle.
order of operations	The conventional order of operations is BIDMAS: brackets first, then powers, then division and multiplication, then addition and subtraction.
outcome	An outcome is the result of a trial.
p(n)	p(n) stands for the probability of event n occurring.
parallel	Two lines that always stay the same distance apart are parallel. Parallel lines never cross or meet.

partition, part	To partition means to split a number into smaller amounts, or parts. For example, 57 could be partitioned into 50 + 7, or 40 + 17.
percentage (%)	A percentage is a fraction expressed as the number of parts per hundred.
perimeter	The perimeter of a shape is the distance around it. It is the total length of the edges.
perpendicular	Two lines are perpendicular to each other if they meet at a right angle.
perpendicular bisector	The perpendicular bisector of a line crosses the line at right angles and cuts it in half.
pi (π)	In a circle, the ratio of the circumference to the diameter is constant. This ratio is called pi, written π. Good approximations are $\pi = 3.14$ or $\pi = \dfrac{22}{7}$.
pie chart	A circle used to display data. The angle of a sector is proportional to the frequency.
plane of symmetry	A plane of symmetry is a cross-section through a 3D shape that divides the shape into two identical halves. All prisms have at least one plane of symmetry.

polygon: pentagon, hexagon, octagon

A polygon is a closed shape with three or more straight edges.

A pentagon has five sides.

A hexagon has six sides.

An octagon has eight sides.

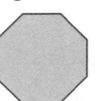

population pyramid	A population pyramid is a back-to-back bar chart comparing two populations.
position-to-term rule	The position-to-term rule links the value of a term to its position in the sequence.

positive	A positive number is greater than zero.
power, index, indices	When a number is written in index notation, the power or index is the raised number. For example, the power of 3^2 is 2.
primary (data)	Data you collect yourself is primary data.
prime factor	A prime factor is a factor that is prime.
prime factor decomposition	Expressing a number as the product of its prime factors is prime factor decomposition. For example, $12 = 2 \times 2 \times 3 = 2^2 \times 3$.
prime number	A prime number is a number that has exactly two different factors, itself and 1.
prism	A prism is a 3D shape with a constant cross-section.

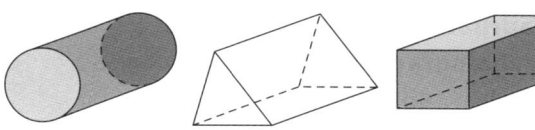

You name a prism by its cross-section.

probability	A measure of how likely an event is.
probability scale	A line numbered 0 to 1 or 0% to 100% on which you place an event based on its probability.
product	The result of a multiplication.
projection: front, side, plan	When you look at a 3D shape from different angles you can see 2D shapes, called projections. The projections of this shape are:

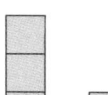

Front   Side   Plan

proportion	Proportion compares the size of a part to the size of a whole. You can express a proportion as a fraction, decimal or percentage.

proportional to	Quantities are proportional to one another when they increase or decrease at the same rate.
pyramid	A pyramid is a 3D shape that tapers to a point called the apex.

 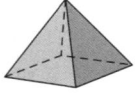 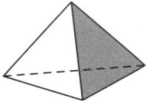

You name a pyramid by its base.

Pythagoras' theorem	In any right-angled triangle, Pythagoras' theorem gives the relationship between the lengths of the sides:

$$a^2 + b^2 = c^2$$

where c is the hypotenuse.

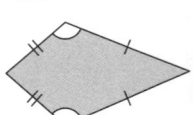

quadrilateral: kite, parallelogram, rectangle, rhombus, square, trapezium

A quadrilateral is a polygon with four sides.

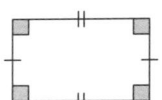

rectangle	parallelogram	kite
All angles are right angles. Opposite sides equal.	Two pairs of parallel sides.	Two pairs of adjacent sides equal. One line of symmetry.

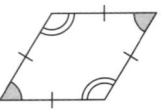 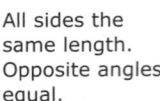 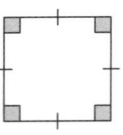

rhombus	square	trapezium
All sides the same length. Opposite angles equal.	All sides and angles equal.	One pair of parallel sides.

questionnaire	A questionnaire is a list of questions used to gather information in a survey.
radius	The radius of a circle is the distance from the centre to the circumference.

random	A selection is random if each object or number is equally likely to be chosen.
range	The range is the difference between the largest and smallest values in a set of data.
ratio	Ratio compares the size of one part with the size of another part.
raw data	Raw data is data that has been collected but not ordered in any way.
rearrange	You rearrange a formula by making the subject a different variable, for example, $A = l \times W$ can be rearranged to $W = \dfrac{A}{l}$.
recurring decimal	A recurring decimal has an unlimited number of digits, which form a repeating pattern, after the decimal point, for example, $\dfrac{1}{3} = 0.333...$
reflect, reflection	A reflection is a transformation in which corresponding points in the object and the image are the same distance from the mirror line. 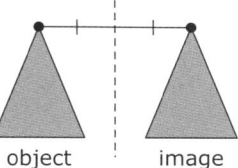
regular polygon	A regular polygon has equal sides and equal angles.
remainder	A remainder is the amount left over when one quantity is divided by another. For example, $9 \div 4 = 2$ remainder 1.
rotate, rotation	A rotation is a transformation in which every point in the object turns through the same angle relative to a fixed point.
rotational symmetry	A shape has rotational symmetry if, when turned, it fits onto itself more than once during a full turn.
round	You round a number by expressing it to a given degree of accuracy. For example, 639 is 600 to the nearest 100 and 12.47 is 12.5 to 1 dp.

rule	A rule describes the link between objects or numbers.
sample	A sample is part of a population.
sample space (diagram)	A sample space diagram records all the outcomes of an experiment.
scale, scale factor	A scale gives the ratio between the size of the object and its diagram. A scale factor is the multiplier in an enlargement.
scale drawing	A scale drawing of an object has every part reduced or enlarged by the same amount, the scale factor.
scatter graph	A scatter graph is a graph on which pairs of observations are plotted.
secondary (data)	Data already collected is secondary data.
sector	Any two radii (plural of radius) will split a circle into two sectors.

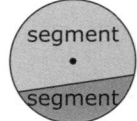

segment	Any line across a circle will split the circle into two segments.

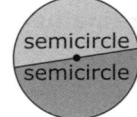

semicircle	A diameter splits a circle into two equal halves called semicircles.
sequence	A set of numbers or diagrams that follow a rule.
similar	Two shapes are similar if the angles are the same and corresponding lengths are in proportion.
simplest form	A fraction (or ratio) is in its simplest form when the numerator and denominator (or parts of the ratio) have no common factors. For example, $\frac{3}{5}$ is expressed in its simplest form.

simplify

To simplify a fraction, you divide the numerator and denominator by their highest common factor. To simplify an expression, you gather all like terms together into a single term.

sine

The sine of an angle in a right-angled triangle is the ratio of the opposite side o to the hypotenuse h.

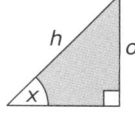

$\sin x = \dfrac{o}{h}$

skew

A distribution is skew if most of the data are at one end of the range.

solid (3D) shape: cube, cuboid, prism, pyramid, square-based pyramid, tetrahedron

A solid is a shape formed in three-dimensional space.

cube

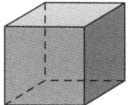

six square faces

cuboid

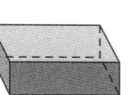

six rectangular faces

prism

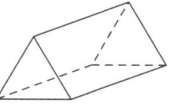

the cross-section is constant

pyramid

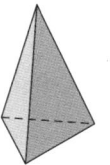

the faces meet at a common vertex

tetrahedron

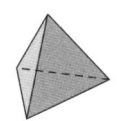

all the faces are equilateral triangles

square-based pyramid

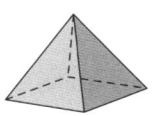

the base is a square

solution (of an equation) The solution of an equation is the value of the variable that makes the equation true.

square number, squared If you multiply a number by itself the result is a square number.

For example, 25 is a square number because $5^2 = 5 \times 5 = 25$.

square root A square root is a number that when multiplied by itself is equal to a given number.

For example $\sqrt{25} = 5$, because $5 \times 5 = 25$.

stem-and-leaf diagram A way of displaying grouped data.

For example, the numbers 29, 16, 18, 8, 4, 16, 27, 19, 13 and 15 could be displayed as:

0	4 8
1	3 5 6 6 8 9
2	7 9

Key: $\boxed{0 \mid 4}$ means 4

straight-line graph When coordinate points lie in a straight line they form a straight-line graph. It is the graph of a linear equation.

subject The subject of a formula is the term on its own on the left of the equals sign.

substitute When you substitute you replace part of an expression with a particular value.

sum The total and is the result of an addition.

surface, surface area The surface area of a solid is the total area of its faces.

survey An investigation to find information.

symmetric, symmetrical A shape is symmetrical if a line divides it into two equal parts. The line is a line of symmetry.

T(n) T(n) is the notation for the general, nth, term of a sequence.

For example, T(3) is the third term.

tangent (of an angle)	The tangent of an angle in a right-angled triangle is the ratio of the opposite side o to the adjacent side a. 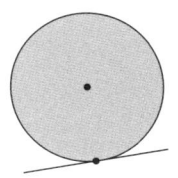 $\tan x = \dfrac{o}{a}$
tangent (to a circle)	A tangent to a circle is a line that touches the circle in one place.

tangent

tenth	A tenth is 1 out of 10 or $\dfrac{1}{10}$. For example, 0.5 has 5 tenths.
term	A term is a number or object in a sequence or part of an expression.
term-to-term rule	A term-to-term rule links a term in a sequence to the previous term.
tessellate, tessellation	A tessellation is a tiling pattern with no gaps.
theoretical probability	A theoretical probability is worked out without an experiment by considering all the possible outcomes.
thousandth	A thousandth is 1 out of 1000 or $\dfrac{1}{1000}$. For example, 0.002 has 2 thousandths.
three-dimensional (3D)	Any solid shape is three-dimensional.
transformation	A transformation moves a shape from one place to another.
translate, translation	A translation is a transformation in which every point in an object moves the same distance and direction. It is a sliding movement.
tree diagram	A diagram showing the possible outcomes of one or more events. You write the outcomes at the end of branches and the probabilities on the branches.

trend	A relationship between observed data and an independent variable such as time.
trial	A single observation in an experiment.
trial and improvement	To find the answer to a complex calculation it is sometimes easier to estimate the answer then improve the estimate. This is called trial and improvement.

triangle: equilateral, isosceles, scalene, right-angled

A triangle is a polygon with three sides.

equilateral

three equal
sides

isosceles

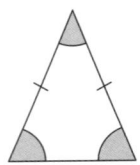

two equal
sides

scalene

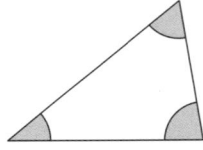

no equal
sides

right-angled

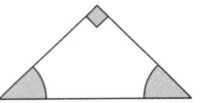

one angle is
90°

triangular number

The triangular numbers form the sequence 1, 3, 6, 10, 15, 21, 28, ... They are the number of dots in a triangular pattern.

two-way table

A two-way table links two independent variables, for example, the result when you toss two dice and add the scores.

		Dice 2					
		1	2	3	4	5	6
Dice 1	1	2	3	4	5	6	7
	2	3	4	5	6	7	8
	3	4	5	6	7	8	9
	4	5	6	7	8	9	10
	5	6	7	8	9	10	11
	6	7	8	9	10	11	12

unit fraction	A unit fraction has 1 as the numerator, for example, $\frac{1}{2}$, $\frac{1}{7}$, $\frac{1}{23}$.
unitary method	In a unitary method you first work out the size of a single unit and then scale it up or down.
unknown	An unknown is a variable. You can often find the value of an unknown by solving an equation.
value	The amount an expression or variable is worth.
variable	A symbol that can take a range of values.
vector	You can use a vector to specify a translation. For example, $\begin{pmatrix} 3 \\ 4 \end{pmatrix}$ means you move the object 3 units to the right and 4 units up. To move left or down you use negative numbers in the vector.
vertex, vertices	A vertex of a solid is a point at which two or more edges meet. A vertex of a 2D shape is where two sides meet.
vertical	Vertical means straight up and down, at right angles to the horizontal.
vertically opposite angles	When two straight lines cross they form two pairs of equal angles called vertically opposite angles. $a = c \qquad b = d$
volume: cubic millimetre, cubic centimetre, cubic metre	The volume of an object is a measure of how much space it occupies.